Justification with Nominals

Inauguraldissertation
der Philosophisch-naturwissenschaftlichen Fakultät
der Universität Bern

vorgelegt von
Alexander Kashev
aus Russland

Leiter der Arbeit:
Prof. Dr. Th. Studer
Institut für Informatik

Von der Philosophisch-naturwissenschaftlichen Fakultät angenommen.

Bern, den 13. Dezember 2016   Der Dekan:
                                                        Prof. Dr. G. Colangelo

© 2016 Alexander Kashev.

This work is licensed under the Creative Commons Attribution-ShareAlike 4.0 International License. To view a copy of this license, visit: http://creativecommons.org/licenses/by-sa/4.0/

ISBN: 978-1-326-85065-4

Acknowledgments

I would like to thank Prof. Dr. Thomas Studer for supervising me during my doctoral studies. His knowledge, advice and, importantly, trust and encouragement were key for my research.

I would also like to thank Prof. Dr. Gerhard Jäger for his continued support and trust, as well as providing an excellent working environment during my studies in Bern.

I would like to thank Dr. Roman Kuznets for presenting me with an opportunity to study in Bern, suggesting the initial idea for research presented in this thesis, and many helpful remarks during the draft stage of this thesis.

I would like to thank all of my colleagues from Logic and Theory Group, who were always there to both help with my work and help relax in-between work. My life in Switzerland wouldn't be the same without them.

I would like to thank my former supervisors in Moscow, Prof. Dr. Lev D. Beklemishev and Dr. Vladimir N. Krupski, for cultivating my love for logic and making possible my studies in Bern.

Last but not least, I would like to thank my friends and my family for their support during my studies, including believing in me when I needed it most.

The research for this thesis was supported by the Swiss National Science Foundation.

Contents

Introduction   1

I. Ground definitions and preexisting work   5

1. Common notational framework   7

2. Justification Logic With Updates: JUP   9
  2.1. Language $\mathcal{L}_s^+$ 9
  2.2. Logic JUP 10
  2.3. Semantics for JUP 12
  2.4. Soundness and completeness for JUP 14

II. Rollback for JUP   15

3. System $\text{JUP}^\pm$ 17
  3.1. Language $\mathcal{L}_s^\pm$ 17
  3.2. Logic $\text{JUP}^\pm$ 18
  3.3. Semantics for $\text{JUP}^\pm$ 20

4. Soundness and completeness for $\text{JUP}^\pm$ 25
  4.1. Soundness for $\text{JUP}^\pm$ 25
  4.2. Completeness for $\text{JUP}^\pm$ 30

5. Rollback properties: history erasing and contraction   43
  5.1. History erasing 43
  5.2. Rollback and contraction 46

III. Nominals — 57

6. From updates to nominals — 59

7. System JN_V — 63
   7.1. Language $\mathcal{L}_n^V$ 63
   7.2. Logic JN_V 66
   7.3. Semantics for JN_V 67

8. Soundness and completeness for JN_V — 73
   8.1. Soundness for JN_V 73
   8.2. Completeness for JN_V 78

9. Finite model property for JN_V — 83

10. Atomic model semantics for JN_V — 91
   10.1. Preliminary definitions and lemmas 92
   10.2. Atomization 100

11. Concluding remarks on nominals — 117

Bibliography — 119

Index — 121

Introduction

Dynamic justification logic

The broad field of *epistemic logic* aims to reason about knowledge and belief. Traditional *epistemic modal logic* deals with the fact of knowing or believing a certain statement: $\Box F$ expresses the fact that an agent knows (believes) statement F, regardless of the underlying reason for that knowledge (belief).

In contrast, the framework of *justification logic* introduced by Artemov [Art01] provides tools to formalize explicit reasons for such belief. Justification terms serve as representation of evidence an agent has for its belief: $t : F$ expresses the fact that an agent believes in statement F for the reason t, or t is agent's evidence for F.

Tracking of explicit evidence allows to trace the reasoning the agent performed to come to a conclusion from base evidence. As an example, the application axiom illustrates the use of evidence when the agent closes its beliefs under modus ponens:

$$t : (F \to G) \wedge s : F \to t \cdot s : G$$

Such a framework makes it possible to analyze many epistemic problems and puzzles from a new perspective [Art06, Art08, BRS14, BKS11a, KMOS15, KOS16].

Traditional justification logic provides a static picture of all justified beliefs that emerge from a given set of premises. It's a natural question to consider an extension for dynamically changing sets of reasons for beliefs, for example as a result of communication; this is studied in the field of *dynamic justification logic*, started by Renne [Ren08] and continued in [BKR+10, BKS11b, BRS12, BKS14, BRS15].

Kuznets and Studer in [KS13] introduced a dynamic justification logic $\mathsf{JUP}_{\mathsf{CS}}$, providing a simple axiomatization for belief expansion and minimal change. It adds a new kind of atomic evidence term, $\mathsf{up}(F)$, representing

the evidence for a formula F after updating the belief set with F, and also explicitly doesn't provide evidence for anything else. It also adds an update operator $[\Gamma]F$, which represents the statement F in the state after an update with a finite set of formulas Γ. The two are connected with the following axiom:
$$[\Gamma]\mathsf{up}(F) : F \quad \text{if } F \in \Gamma$$

$\mathsf{JUP}_{\mathsf{CS}}$ is shown to be sound and complete with respect to a class of basic modular models [Art12, KS12] called *generated models* [Stu13], with the evidence relation generated inductively from an evidence basis as a least fixed point of a closure operator. Furthermore, the bases use only atomic terms.

Being able to restrict the model definition to atomic terms produces simple models to work with, but comes at a price: term application has to carry a record of the formula that was used in the application, e.g. $t \cdot_F s$, and the application axiom was modified to reflect this:

$$t : (F \to G) \wedge s : F \leftrightarrow t \cdot_F s : G$$

Update rollback

Logic $\mathsf{JUP}_{\mathsf{CS}}$ provide syntax and semantics for updates: an operation that adds an explicit reason $\mathsf{up}(F)$ to believe a certain statement F. A natural question is to consider the inverse of the update operation: taking away an explicit reason to believe a statement F.

In the simplest form, this operation would only remove specific reason $\mathsf{up}(F)$ instead of arbitrary justification terms, essentially allowing to undo updates without disturbing initial beliefs.

Borrowing the terminology for undoing software updates and patches, we called this operation *rollback*.

To provide axiomatization for rollback, we add a new operator, $[\Gamma^\circ]F$, representing the statement F in the state after rolling back updates with a finite set of formulas Γ.

By extending definitions and proofs from [KS13] to encompass the new operator, we obtain a similar semantics based on generated models with atomic evidence bases.

We further clarify the intended semantics for a rollback following a chain of updates and rollbacks with the help of *history erasing* theorem.

Finally, we explore the connections between rollbacks and belief contraction [AGM85]. We demonstrate that a couple of naive contraction operators based on rollbacks fail to meet the requirements of belief contraction, but show that a certain generalization is, for well-behaved models, a candidate for shielded belief contraction [FH01], which opens further opportunities for research.

Nominals

Term application with a subscript is an uncommon addition to justification logic, first introduced by Renne [Ren09]. It naturally raised the question whether it's possible to remove the subscript and use the more traditional application axiom for justification logic for dynamic justification with updates.

Naively modified axioms lead to semantics of generated model with non-atomic bases, where you can have compound justification terms that justify some statement, but constituent parts of that term do not justify anything. Besides not being very natural, this also stops the original completeness proof from working, as it essentially relied on the ability to reduce justification by application terms to justification by subterms.

This left the question whether it's possible to remove the subscript while maintaining semantics with atomic models. This thesis presents a stepping stone in the study of this question. To examine the most basic form of justification arising from updates, we look at the situation after a finite set of updates with atomic propositional statements, omitting the dynamics of updates. Namely, for propositional variables in this fixed update set, we introduce *nominals*: terms that justify exactly one formula, which is enforced on axiom level.

An axiom system $\mathsf{JN}_{\mathsf{CS}}^{\mathsf{V}}$ is formulated to capture that property, and shown to be sound and complete with respect to a class of generated models called *intermediate models*. This semantics is then shown to enjoy a variant of finite model property.

However, this class of models is still less natural than the original semantics for $\mathsf{JUP}_{\mathsf{CS}}$. We show that, for appropriate constant specifications, it's possible to obtain a completeness result for a more natural class of atomic models through a model reduction procedure we call *atomization*.

Those results represent the first step in adapting $\mathsf{JUP}_{\mathsf{CS}}$ to subscript-

Contents

free application. The primary direction for further work is re-introducing dynamic updates and studying the epistemic properties of the resulting system. Another possible direction is extending nominals to non-atomic propositions.

Part I.

Ground definitions and preexisting work

Part I

Ground definitions and prevailing work

1. Common notational framework

First, we will introduce common syntactic conventions that apply to all frameworks discussed.

For every justification logic framework we introduce, its language $\mathcal{L}$ will have a common structure:

- A sort of objects called *terms*, given by a set $\mathsf{Tm}_\mathcal{L}$ and a subset of *atomic terms* $\mathsf{ATm}_\mathcal{L}$.

  We will use syntactic variables $s, t, p, q, r, \ldots$ to denote terms.

- A sort of objects called *formulas*, given by a set $\mathsf{Fm}_\mathcal{L}$ and a subset of *atomic formulas* which we denote $\mathsf{Prop}_\mathcal{L}$, since in all cases it will be the set of propositional variables.

  We will use syntactic variables $F, G, H, \ldots$ to denote formulas, and syntactic variables $P, Q, R, \ldots$ to denote propositional variables.

As we require finite sets of formulas in our syntax, we will use syntactic variables Γ, Δ to denote them.

Terms and formulas are given by a set of inductive (sometimes mutually inductive) rules with atomic terms and formulas explicitly given as a basis for inductive build-up.

Whenever it is clear from the context, we will drop the subscript $\mathcal{L}$ from those sets.

We will use the standard formula connectives explicitly: $\neg F$ for negation and $(F \to G)$ for implication, for $F, G \in \mathsf{Fm}_\mathcal{L}$.

We also use other standard propositional connectives, namely $(F \wedge G)$ for conjunction, $(F \vee G)$ for disjunction and $F \leftrightarrow G$ for equivalence, as implicit shorthand:

$$(F \wedge G) := \neg(F \to \neg G)$$
$$(F \vee G) := (\neg F \to G)$$
$$(F \leftrightarrow G) := (F \to G) \wedge (G \to F)$$

1. Common notational framework

We will omit parentheses for formula connectives when this is unambiguous; for this, we consider implication is right-associative:

$$F \to G \to H = (F \to (G \to H))$$

We also consider that $\wedge$ and $\vee$ bind stronger than $\to$ and $\leftrightarrow$.

Another standard formula connective we will use in every language introduced is evidence: $t : F$ for $t \in \mathsf{Tm}_\mathcal{L}, F \in \mathsf{Fm}_\mathcal{L}$.

Formulas of the form $t : F$ should be read as *"t justifies F"* or *"F is believed for the reason t"*.

Definition 1.1 (Evidence pairs). Given a justification language $\mathcal{L}$, we call pairs of the form $(t, F) \in \mathsf{Tm}_\mathcal{L} \times \mathsf{Fm}_\mathcal{L}$ *evidence pairs*.

2. Justification Logic With Updates: JUP

In this chapter, we present the logic JUP and corresponding semantics from [KS13]. We will also formulate the soundness and completeness results for them.

2.1. Language $\mathcal{L}_s^+$

The language of logic JUP, which we will denote $\mathcal{L}_s^+$ and is presented in this chapter, differs from the standard language of justification logic in two ways:

- Term application carries a formula subscript: $t \cdot_F s$.
- An additional constructor clause for formulas is introduced: $[\Gamma]F$ for a formula F and a finite set of formulas Γ.

Having a formula subscript means that terms and formulas have to be defined by mutual induction:

Definition 2.1 (Terms and formulas of $\mathcal{L}_s^+$). The sets $\mathsf{Tm}_{\mathcal{L}_s^+}$ of *terms* and $\mathsf{Fm}_{\mathcal{L}_s^+}$ of *formulas* are defined inductively by the following clauses:

Terms:

- (Countably many) term variables: $\{x, \ldots\} \subseteq \mathsf{Tm}_{\mathcal{L}_s^+}$
- (Countably many) term constants: $\{c, \ldots\} \subseteq \mathsf{Tm}_{\mathcal{L}_s^+}$
- Term application: $(t \cdot_F s) \in \mathsf{Tm}_{\mathcal{L}_s^+}$ for $t, s \in \mathsf{Tm}_{\mathcal{L}_s^+}$, $F \in \mathsf{Fm}_{\mathcal{L}_s^+}$
- Update terms: $\mathsf{up}(F) \in \mathsf{Tm}_{\mathcal{L}_s^+}$ for $F \in \mathsf{Fm}_{\mathcal{L}_s^+}$

Formulas:

- (Countably many) propositional variables: $\{P,\ldots\} \subseteq \mathsf{Fm}_{\mathcal{L}_s^+}$
- Implication: $(F \to G) \in \mathsf{Fm}_{\mathcal{L}_s^+}$ for $F, G \in \mathsf{Fm}_{\mathcal{L}_s^+}$
- Negation: $\neg F \in \mathsf{Fm}_{\mathcal{L}_s^+}$ for $F \in \mathsf{Fm}_{\mathcal{L}_s^+}$
- Evidence: $t : F \in \mathsf{Fm}_{\mathcal{L}_s^+}$ for $t \in \mathsf{Tm}_{\mathcal{L}_s^+}, F \in \mathsf{Fm}_{\mathcal{L}_s^+}$
- Update: $[\Gamma]F \in \mathsf{Fm}_{\mathcal{L}_s^+}$ for $F \in \mathsf{Fm}_{\mathcal{L}_s^+}$ and finite $\Gamma \subseteq \mathsf{Fm}_{\mathcal{L}_s^+}$

Term variables, term constants and update terms are collectively called *atomic terms*, and their set is denoted by $\mathsf{ATm}_{\mathcal{L}_s^+}$.

Propositional variables are called *atomic formulas*, and their set is denoted by $\mathsf{Prop}_{\mathcal{L}_s^+}$.

The intended meaning of formulas of the form $[\Gamma]F$ is *"F holds after an update with all the formulas in Γ"*, and $\mathsf{up}(F)$ represents the special evidence for a formula F in case there was an update including F.

We will omit parentheses for term application when this is unambiguous; for this, we consider implication is left-associative:

$$t \cdot_F s \cdot_G r = ((t \cdot_F s) \cdot_G r)$$

We will also drop the subscript $\mathcal{L}_s^+$ for the rest of this chapter.

2.2. Logic JUP

To define the deduction system $\mathsf{JUP}_{\mathsf{CS}}$, we first give a list of axiom schemes, using shorthand notation introduced in Chapter 1:

Definition 2.2 (Axioms of JUP).

1. All classical propositional tautologies (Taut)
2. $t : (F \to G) \land s : F \leftrightarrow t \cdot_F s : G$ (App)
3. $[\Gamma]P \leftrightarrow P \qquad P \in \mathsf{Prop}$ (Red.1)
4. $[\Gamma]\neg F \leftrightarrow \neg[\Gamma]F$ (Red.2)
5. $[\Gamma](F \to G) \leftrightarrow ([\Gamma]F \to [\Gamma]G)$ (Red.3)

2.2. Logic JUP

6. $t : F \to [\Gamma]t : F$ (Pers)
7. $\neg \mathsf{up}(F) : G$ (Init)
8. $[\Gamma]\mathsf{up}(F) : F$ if $F \in \Gamma$ (Up)
9. $[\Gamma]t : F \to t : F$
      if $t \in \mathsf{ATm}$ and either $t \neq \mathsf{up}(F)$ or $F \notin \Gamma$ (MC.1)
10. $[\Gamma]t \cdot_F s : G \leftrightarrow [\Gamma]t : (F \to G) \wedge [\Gamma]s : F$ (MC.2)
11. $[\Gamma][\Delta]F \leftrightarrow [\Gamma \cup \Delta]F$ (It)

Then, we define the notion of constant specification, which parametrizes the deduction system:

Definition 2.3. A *constant specification* CS is a set of evidence pairs of the form $(c, c_1 : \ldots : c_n : F)$ (including $n = 0$, i.e. (c, F)), where F is an axiom instance of JUP and $c, c_1, \ldots, c_n$ are term constants.

For a given constant specification CS, we can define the deduction system $\mathsf{JUP_{CS}}$:

Definition 2.4 (Logic $\mathsf{JUP_{CS}}$). The logic $\mathsf{JUP_{CS}}$ is a Hilbert-style deduction system with axiom schemes JUP and two inference rules:

$$\frac{F \quad F \to G}{G} \text{ (MP)}$$

$$\frac{(c, F) \in \mathsf{CS}}{c : F} \text{ (AN)}$$

Remarks on $\mathsf{JUP_{CS}}$:

- The axiom (App) differs from the standard justification logic axiom of application (e.g. [Art01]) — the implication goes both ways, allowing to unambiguously infer premises from the conclusion. This is made possible by retaining the formula used in application premises in the subscript of the application.

- The group of axioms (Red.1)—(Red.3) are called *reduction axioms* and allow for carrying updates through propositional connectives until they apply directly to evidence formulas: $[\Gamma]t : F$.

- The axiom (Pers) is called *persistence axiom* and reflects the fact that updates only add potential evidence, but existing evidence remains valid.

- The axioms (Init) and (Up) deal with the update terms. (Init) reflects the fact that without updates, no update term can be considered evidence for any formula. (Up) reflects that after an update that contains a formula F, the corresponding evidence term $\mathsf{up}(F)$ becomes valid evidence for it.

- The group of axioms (MC.1) and (MC.2) reflect the principle of *minimal change*: together, they show that updates do not matter for terms that do not contain relevant update terms as subterms, and that the effect of the updates reduces to effect on the constituent atomic terms. Again, this requires the subscript in application to formulate (MC.2).

- The last axiom, (It), shows that *iterated* updates can always be simplified to a single combined update, and vice versa.

- The constant specification CS reflects the ability to use axioms to construct a justified formula $t : F$ from premises of same form, using axiom (App) as an analogue of modus ponens. Different constant specifications yield different sets of justifiable formulas and different possible justifications for them.

Definition 2.5. A JUP formula F is called *provable in* $\mathsf{JUP}_{\mathsf{CS}}$ if it can be derived from axioms and rules of $\mathsf{JUP}_{\mathsf{CS}}$.

It is denoted as $\vdash_{\mathsf{JUP}_{\mathsf{CS}}} F$.

2.3. Semantics for JUP

To provide semantics for $\mathsf{JUP}_{\mathsf{CS}}$, [KS13] uses a special class of basic modular models [Art12, KS12] called *generated models* [Stu13], in which the evidence relation is built up from a limited basis using a least fixpoint construction.

Definition 2.6 (Atomic basis). An *atomic basis* is an arbitrary set of evidence pairs with atomic term part: $\mathcal{B} \subseteq \mathsf{ATm} \times \mathsf{Fm}$.

2.3. Semantics for JUP

We will use the syntactic variable $\mathcal{B}$ to denote bases.

Then, we define an operator on the powerset $\mathcal{P}(\mathsf{Tm} \times \mathsf{Fm})$, parametrized by a basis:

Definition 2.7 (Evidence closure operator). For a set $X \subseteq \mathsf{Tm} \times \mathsf{Fm}$ and an atomic basis $\mathcal{B} \subseteq \mathsf{ATm} \times \mathsf{Fm}$ define an operator $\mathsf{cl}_\mathcal{B}^{\mathsf{JUP}}(X)$ by clauses:

- $(t, F) \in \mathcal{B} \Rightarrow (t, F) \in \mathsf{cl}_\mathcal{B}^{\mathsf{JUP}}(X)$ (equivalently, $\mathcal{B} \subseteq \mathsf{cl}_\mathcal{B}^{\mathsf{JUP}}(X)$)
- $(t, F \to G) \in X, (s, F) \in X \Rightarrow (t \cdot_F s, G) \in \mathsf{cl}_\mathcal{B}^{\mathsf{JUP}}(X)$

This is a monotone operator on $\mathcal{P}(\mathsf{Tm} \times \mathsf{Fm})$: for every $X, Y \subseteq \mathsf{Tm} \times \mathsf{Fm}$,

$$X \subseteq Y \Rightarrow \mathsf{cl}_\mathcal{B}^{\mathsf{JUP}}(X) \subseteq \mathsf{cl}_\mathcal{B}^{\mathsf{JUP}}(Y)$$

Therefore, by Knaster–Tarski theorem [Tar55] it has a least fixpoint, which will define our evidence relation:

Definition 2.8 (Evidence relation). For an atomic basis $\mathcal{B} \subseteq \mathsf{ATm} \times \mathsf{Fm}$, define the (minimal) evidence relation $\mathcal{E}^{\mathsf{JUP}}(\mathcal{B})$ as the l.f.p. of $\mathsf{cl}_\mathcal{B}^{\mathsf{JUP}}$.

The final ingredient to define a generated model is a valuation of propositional variables:

Definition 2.9 (Propositional valuation). A *propositional valuation* is an arbitrary subset of propositional variables $\mathsf{v} \subseteq \mathsf{Prop}$.

Definition 2.10 (JUP models). A JUP *model* is a pair $\mathcal{M} = (\mathsf{v}, \mathcal{B})$ with propositional valuation $\mathsf{v} \subseteq \mathsf{Prop}$ and atomic basis $\mathcal{B} \subseteq \mathsf{ATm} \times \mathsf{Fm}$.

Before we can interpret JUP formulas in the semantics, we need to introduce model updates.

Definition 2.11 (Update set). For a finite set of formulas Γ, define the *update set* $\mathcal{U}_\Gamma$:

$$\mathcal{U}_\Gamma := \{(\mathsf{up}(F), F) \mid F \in \Gamma\}$$

Definition 2.12 (Updated model). For a JUP model $\mathcal{M} = (\mathsf{v}, \mathcal{B})$ and a finite set of formulas Γ, define the *updated model* $\mathcal{M}^{+\Gamma} = (\mathsf{v}, \mathcal{B}^{+\Gamma})$, where $\mathcal{B}^{+\Gamma} := \mathcal{B} \cup \mathcal{U}_\Gamma$.

Now we can interpret formulas in a model, using the truth relation:

2. Justification Logic With Updates: JUP

Definition 2.13 (JUP truth relation). For a JUP model $\mathcal{M} = (v, \mathcal{B})$, and a formula F, the *truth relation* $\mathcal{M} \Vdash F$ is defined inductively by:

- $\mathcal{M} \Vdash P \Leftrightarrow P \in v$
- $\mathcal{M} \Vdash \neg F \Leftrightarrow \mathcal{M} \nVdash F$
- $\mathcal{M} \Vdash F \to G \Leftrightarrow (\mathcal{M} \nVdash F \text{ or } \mathcal{M} \Vdash G)$
- $\mathcal{M} \Vdash t : F \Leftrightarrow (t, F) \in \mathcal{E}^{\mathsf{JUP}}(\mathcal{B})$
- $\mathcal{M} \Vdash [\Gamma] F \Leftrightarrow \mathcal{M}^{+\Gamma} \Vdash F$

If $\mathcal{M} \Vdash F$, we say that F *is true in* $\mathcal{M}$.

We will drop the superscript JUP in $\mathcal{E}^{\mathsf{JUP}}$ for the rest of this chapter.

2.4. Soundness and completeness for JUP

[KS13] shows that $\mathsf{JUP}_{\mathsf{CS}}$ is sound and complete with respect to a certain class of JUP models. We will only formulate the results here; proofs in the next part follow the same shape as the original proofs.

Definition 2.14 (CS-model). For a constant specification CS, a JUP model $\mathcal{M} = (v, \mathcal{B})$ is called a CS-*model* if $\mathsf{CS} \subseteq \mathcal{B}$.

Definition 2.15 (Initial model). JUP model $\mathcal{M} = (v, \mathcal{B})$ is called *initial* if, for all formulas F, G, $(\mathsf{up}(F), G) \notin \mathcal{B}$.

Essentially, initial models represent the state before any updates are applied, and updated models reflect the changes to the state brought on by updates.

Definition 2.16 (Validity). A JUP formula F is called *valid w.r.t. initial* CS-*models* if, for every initial CS-model $\mathcal{M}$, we have $\mathcal{M} \Vdash F$.

It is denoted as $\Vdash_{\mathsf{JUP}_{\mathsf{CS}}} F$.

Theorem 2.17 (Soundness and completeness of $\mathsf{JUP}_{\mathsf{CS}}$). *For a given constant specification* CS, *a* JUP *formula* F *is valid w.r.t. initial* CS-*models iff it is provable in* $\mathsf{JUP}_{\mathsf{CS}}$:

$$\Vdash_{\mathsf{JUP}_{\mathsf{CS}}} F \Leftrightarrow \vdash_{\mathsf{JUP}_{\mathsf{CS}}} F$$

Part II.

Rollback for JUP

3. System JUP$^{\pm}$

The goal of this chapter is to extend JUP with a new operator, *rollback*, that cancels previous updates.

3.1. Language $\mathcal{L}_s^{\pm}$

The language that will be used in this chapter is an extension of the language $\mathcal{L}_s^+$ introduced in Section 2.1.

Definition 3.1 (Terms and formulas of $\mathcal{L}_s^{\pm}$)**.** The sets $\mathsf{Tm}_{\mathcal{L}_s^{\pm}}$ of *terms* and $\mathsf{Fm}_{\mathcal{L}_s^{\pm}}$ of *formulas* are defined inductively by the following clauses:

Terms:

- (Countably many) term variables: $\{x, \ldots\} \subseteq \mathsf{Tm}_{\mathcal{L}_s^{\pm}}$
- (Countably many) term constants: $\{c, \ldots\} \subseteq \mathsf{Tm}_{\mathcal{L}_s^{\pm}}$
- Term application: $(t \cdot_F s) \in \mathsf{Tm}_{\mathcal{L}_s^{\pm}}$ for $t, s \in \mathsf{Tm}_{\mathcal{L}_s^{\pm}}$, $F \in \mathsf{Fm}_{\mathcal{L}_s^{\pm}}$
- Update terms: $\mathsf{up}(F) \in \mathsf{Tm}_{\mathcal{L}_s^{\pm}}$ for $F \in \mathsf{Fm}_{\mathcal{L}_s^{\pm}}$

Formulas:

- (Countably many) propositional variables: $\{P, \ldots\} \subseteq \mathsf{Fm}_{\mathcal{L}_s^{\pm}}$
- Implication: $(F \to G) \in \mathsf{Fm}_{\mathcal{L}_s^{\pm}}$ for $F, G \in \mathsf{Fm}_{\mathcal{L}_s^{\pm}}$
- Negation: $\neg F \in \mathsf{Fm}_{\mathcal{L}_s^{\pm}}$ for $F \in \mathsf{Fm}_{\mathcal{L}_s^{\pm}}$
- Evidence: $t : F \in \mathsf{Fm}_{\mathcal{L}_s^{\pm}}$ for $t \in \mathsf{Tm}_{\mathcal{L}_s^{\pm}}$, $F \in \mathsf{Fm}_{\mathcal{L}_s^{\pm}}$
- Update: $[\Gamma]F \in \mathsf{Fm}_{\mathcal{L}_s^{\pm}}$ for $F \in \mathsf{Fm}_{\mathcal{L}_s^{\pm}}$ and finite $\Gamma \subseteq \mathsf{Fm}_{\mathcal{L}_s^{\pm}}$
- Rollback: $[\Gamma^{\circ}]F$ for $F \in \mathsf{Fm}_{\mathcal{L}_s^{\pm}}$ and finite $\Gamma \subseteq \mathsf{Fm}_{\mathcal{L}_s^{\pm}}$

3. System JUP$^{\pm}$

Term variables, term constants and update terms are collectively called *atomic terms*, and their set is denoted by $\mathsf{ATm}_{\mathcal{L}_s^{\pm}}$.

Propositional variables are called *atomic formulas*, and their set is denoted by $\mathsf{Prop}_{\mathcal{L}_s^{\pm}}$.

As before, term application is considered left-associative, and we will drop the subscript $\mathcal{L}_s^{\pm}$ for the rest of Part II.

The intended meaning of formulas of the form $[\Gamma^{\circ}]F$ is *"F holds after a rollback of all the formulas in Γ"*, which means invalidating all previous updates, if any, for formulas in Γ.

One important thing to note is the intended ordering of iterated updates and rollbacks. It did not matter for JUP, since the axiom (It) makes updates commutative, however from the semantics presented in Section 2.3 it is clear that $[\Gamma][\Delta]F$ is interpreted as *"after an update with Γ, and subsequently an update with Δ, F holds"*.

As such, we interpret iterated updates and rollbacks in a formula as a sequence from left to right.

3.2. Logic JUP$^{\pm}$

The deduction system $\mathsf{JUP}_{\mathsf{CS}}$ is an extension of the deduction system $\mathsf{JUP}_{\mathsf{CS}}$ introduced in Section 2.2. It uses shorthand notation from Chapter 1 for propositional connectives:

Definition 3.2 (Axioms of JUP$^{\pm}$).

1. All classical propositional tautologies (Taut)
2. $t : (F \to G) \land s : F \leftrightarrow t \cdot_F s : G$ (App)
3. $[\Gamma]P \leftrightarrow P \qquad P \in \mathsf{Prop}$ (Red.1)
4. $[\Gamma]\neg F \leftrightarrow \neg[\Gamma]F$ (Red.2)
5. $[\Gamma](F \to G) \leftrightarrow ([\Gamma]F \to [\Gamma]G)$ (Red.3)
6. $t : F \to [\Gamma]t : F$ (Pers)
7. $\neg\mathsf{up}(F) : G$ (Init)
8. $[\Gamma]\mathsf{up}(F) : F \qquad$ if $F \in \Gamma$ (Up)
9. $[\Gamma]t : F \to t : F$
    if $t \in \mathsf{ATm}$ and either $t \neq \mathsf{up}(F)$ or $F \notin \Gamma$ (MC.1)

3.2. Logic JUP$^\pm$

10. $[\Gamma]t \cdot_F s : G \leftrightarrow [\Gamma]t : (F \to G) \wedge [\Gamma]s : F$ (MC.2)
11. $[\Gamma][\Delta]F \leftrightarrow [\Gamma \cup \Delta]F$ (It)
12. $[\Gamma^\circ]F \leftrightarrow F$ (Roll)
13. $[\Gamma][\Delta^\circ]F \leftrightarrow [\Gamma \setminus \Delta]F$ (Int)

Definition 3.3. A *constant specification* CS is a set of evidence pairs of the form $(c, c_1 : \ldots : c_n : F)$ (including $n = 0$, i.e. (c, F)), where F is an axiom instance of JUP$^\pm$ and $c, c_1, \ldots, c_n$ are term constants.

For a given constant specification CS, we can define the deduction system JUP$^\pm_{\mathsf{CS}}$:

Definition 3.4 (Logic JUP$^\pm_{\mathsf{CS}}$)**.** The logic JUP$^\pm_{\mathsf{CS}}$ is a Hilbert-style deduction system with axiom schemes JUP$^\pm$ and two inference rules:

$$\frac{F \quad F \to G}{G} \text{ (MP)}$$

$$\frac{(c, F) \in \mathsf{CS}}{c : F} \text{ (AN)}$$

Remarks on JUP$^\pm_{\mathsf{CS}}$:

- The axioms (App)—(It) are exactly the axioms of JUP (Definition 2.2) and do not feature the new construct $[\Gamma^\circ]F$.

- The axiom (Roll) reflects the fact that rolling back updates before any updates are made does not change validity of any formula.

- The axiom (Int) describes the interaction between updates and rollbacks: an update followed by a rollback is equivalent to an update without the formulas that were rolled back.

Definition 3.5. A JUP$^\pm$ formula F is called *provable in* JUP$^\pm_{\mathsf{CS}}$ if it can be derived from axioms and rules of JUP$^\pm_{\mathsf{CS}}$.
It is denoted as $\vdash_{\mathsf{JUP}^\pm_{\mathsf{CS}}} F$.

3.3. Semantics for $\mathsf{JUP}^{\pm}$

Semantics for $\mathsf{JUP}^{\pm}$ is an extension of semantics for JUP presented in Section 2.3, so base definitions that do not differ from JUP models:

Definition 3.6 (Atomic basis). An *atomic basis* is an arbitrary set of evidence pairs with atomic term part: $\mathcal{B} \subseteq \mathsf{ATm} \times \mathsf{Fm}$.

Definition 3.7 (Evidence closure). Take an atomic basis $\mathcal{B} \subseteq \mathsf{ATm} \times \mathsf{Fm}$. For a set $X \subseteq \mathsf{Tm} \times \mathsf{Fm}$ define an operator $\mathsf{cl}_\mathcal{B}^{\mathsf{JUP}^{\pm}}(X)$ by clauses:

- $(t, A) \in \mathcal{B} \Rightarrow (t, A) \in \mathsf{cl}_\mathcal{B}^{\mathsf{JUP}^{\pm}}(X)$ (equivalently, $\mathcal{B} \subseteq \mathsf{cl}_\mathcal{B}^{\mathsf{JUP}^{\pm}}(X)$)
- $(t, A \to B) \in X, (s, A) \in X \Rightarrow (t \cdot_A s, B) \in \mathsf{cl}_\mathcal{B}^{\mathsf{JUP}^{\pm}}(X)$

Note that $\mathsf{cl}_\mathcal{B}^{\mathsf{JUP}^{\pm}}$ is a monotone operator on $\mathcal{P}(\mathsf{Tm} \times \mathsf{Fm})$ and therefore has a least fixpoint by Knaster–Tarski theorem [Tar55].

Definition 3.8 (Evidence relation). For an atomic basis $\mathcal{B} \subseteq \mathsf{ATm} \times \mathsf{Fm}$, define the (minimal) evidence relation $\mathcal{E}^{\mathsf{JUP}^{\pm}}(\mathcal{B})$ as the l.f.p. of $\mathsf{cl}_\mathcal{B}^{\mathsf{JUP}^{\pm}}$.

We will drop the superscript $\mathsf{JUP}^{\pm}$ from $\mathcal{E}^{\mathsf{JUP}^{\pm}}$ for the rest of Part II.

Lemma 3.9 (Evidence properties). *Evidence relation $\mathcal{E}(\mathcal{B})$ has the following properties:*

(a) $\mathcal{B} \subseteq \mathcal{E}(\mathcal{B})$

(b) $(t, F) \in \mathcal{E}(\mathcal{B}) \Rightarrow (t, F) \in \mathcal{B}$ for $t \in \mathsf{ATm}$

(c) $\{(t, F \to G), (s, F)\} \subseteq \mathcal{E}(\mathcal{B}) \Leftrightarrow (t \cdot_F s, G) \in \mathcal{E}(\mathcal{B})$

Proof. Remember that $\mathcal{E}(\mathcal{B})$ is a fixpoint of $\mathsf{cl}_\mathcal{B}^{\mathsf{JUP}^{\pm}}$:

$$\mathsf{cl}_\mathcal{B}^{\mathsf{JUP}^{\pm}}(\mathcal{E}(\mathcal{B})) = \mathcal{E}(\mathcal{B})$$

From this

(a) $\mathcal{B} \subseteq \mathsf{cl}_\mathcal{B}^{\mathsf{JUP}^{\pm}}(X)$ for any X by definition.
In particular, $\mathcal{B} \subseteq \mathsf{cl}_\mathcal{B}^{\mathsf{JUP}^{\pm}}(\mathcal{E}(\mathcal{B})) = \mathcal{E}(\mathcal{B})$.

(b) By definition of $\text{cl}_\mathcal{B}^{\text{JUP}^\pm}$, $\text{cl}_\mathcal{B}^{\text{JUP}^\pm}(\mathcal{E}(\mathcal{B}))$ consists of a union of two sets that correspond to clauses.

Any evidence pair coming from the second clause contains an application and therefore cannot be atomic. Therefore, if $(t, F) \in \mathcal{E}(\mathcal{B})$, it must be included in $\text{cl}_\mathcal{B}^{\text{JUP}^\pm}(\mathcal{E}(\mathcal{B}))$ by first clause, which means $(t, F) \in \mathcal{B}$.

(c) First, assume $\{(t, F \to G), (s, F)\} \subseteq \mathcal{E}(\mathcal{B})$. In that case, by second clause in definition of the closure operator we have

$$(t \cdot_F s, G) \in \text{cl}_\mathcal{B}^{\text{JUP}^\pm}(\mathcal{E}(\mathcal{B})) = \mathcal{E}(\mathcal{B}).$$

Conversely, assume $(t \cdot_F s, G) \in \mathcal{E}(\mathcal{B}) = \text{cl}_\mathcal{B}^{\text{JUP}^\pm}(\mathcal{E}(\mathcal{B}))$.

Since $t \cdot_F s$ is not atomic, it cannot come from the first clause in the definition. Therefore, it comes from the second clause and

$$\{(t, F \to G), (s, F)\} \subseteq \mathcal{E}(\mathcal{B})$$

$\square$

Lemma 3.10 (Evidence monotonicity). *Evidence relation $\mathcal{E}(\mathcal{B})$ is monotone on $\mathcal{B}$: for all bases $\mathcal{B}, \mathcal{B}'$ we have*

$$\mathcal{B} \subseteq \mathcal{B}' \Rightarrow \mathcal{E}(\mathcal{B}) \subseteq \mathcal{E}(\mathcal{B}')$$

Proof. Assume $\mathcal{B} \subseteq \mathcal{B}'$ and $(t, F) \in \mathcal{E}(\mathcal{B})$.
We prove $(t, F) \in \mathcal{E}(\mathcal{B}')$ by induction on t.
Base: $t \in \text{ATm}$. In this case, by Lemma 3.9,

$$(t, F) \in \mathcal{E}(\mathcal{B}) \Rightarrow (t, F) \in \mathcal{B} \subseteq \mathcal{B}' \subseteq \mathcal{E}(\mathcal{B}')$$

Step: $t = r \cdot_G s$. Using Lemma 3.9,

$$(r \cdot_G s, F) \in \mathcal{E}(\mathcal{B}) \overset{(3.9)}{\Longrightarrow} ((r, G \to F) \in \mathcal{E}(\mathcal{B}) \text{ and } (s, G) \in \mathcal{E}(\mathcal{B}))$$
$$\overset{\text{(IH)}}{\Longrightarrow} ((r, G \to F) \in \mathcal{E}(\mathcal{B}') \text{ and } (s, G) \in \mathcal{E}(\mathcal{B}'))$$
$$\overset{(3.9)}{\Longrightarrow} (r \cdot_G s, F) \in \mathcal{E}(\mathcal{B}')$$

$\square$

3. System JUP$^\pm$

Definition 3.11 (Models). A JUP$^\pm$ *model* is a pair $\mathcal{M} = (\mathsf{v}, \mathcal{B})$ with propositional valuation $\mathsf{v} \subseteq \mathsf{Prop}$ and atomic basis $\mathcal{B} \subseteq \mathsf{ATm} \times \mathsf{Fm}$.

So far, the models are identical to JUP. However, to interpret rollbacks we need to add a notion of rolled-back models.

Definition 3.12 (Update set). For a finite set of formulas Γ, define the *update set* $\mathcal{U}_\Gamma$:
$$\mathcal{U}_\Gamma := \{(\mathsf{up}(F), F) \mid F \in \Gamma\}$$

Definition 3.13 (Updated model, rolled-back model).
For a model $\mathcal{M} = (\mathsf{v}, \mathcal{B})$, define:

- The *updated model* $\mathcal{M}^{+\Gamma} = (\mathsf{v}, \mathcal{B}^{+\Gamma})$, where $\mathcal{B}^{+\Gamma} := \mathcal{B} \cup \mathcal{U}_\Gamma$.

- The *rolled-back model* $\mathcal{M}^{-\Gamma} = (\mathsf{v}, \mathcal{B}^{-\Gamma})$, where $\mathcal{B}^{-\Gamma} := \mathcal{B} \setminus \mathcal{U}_\Gamma$.

Let σ, σ' stand for either "+" or "−", and let $\mathcal{M} = (\mathsf{v}, \mathcal{B})$ be an arbitrary JUP$^\pm$ model.

Then, for a singleton set of formulas $\Gamma = \{A\}$ we write $\mathcal{M}^{\sigma A}$ instead of $\mathcal{M}^{\sigma\{A\}}$, and for any finite sets of formulas Γ, Δ we abbreviate $(\mathcal{M}^{\sigma\Gamma})^{\sigma'\Delta}$ as $\mathcal{M}^{\sigma\Gamma\sigma'\Delta}$, and likewise for $(\mathcal{B}^{\sigma\Gamma})^{\sigma'\Delta}$.

Using the rolled-back models, we can extend the Definition 2.13 of truth in JUP to account for rollbacks:

Definition 3.14 (Truth). For a JUP$^\pm$ model $\mathcal{M} = (\mathsf{v}, \mathcal{B})$, and a JUP$^\pm$ formula F, the relation $\mathcal{M} \Vdash F$ is defined inductively by:

- $\mathcal{M} \Vdash P \Leftrightarrow P \in \mathsf{v}$

- $\mathcal{M} \Vdash \neg F \Leftrightarrow \mathcal{M} \nVdash F$

- $\mathcal{M} \Vdash F \rightarrow G \Leftrightarrow (\mathcal{M} \nVdash F \text{ or } \mathcal{M} \Vdash G)$

- $\mathcal{M} \Vdash t : F \Leftrightarrow (t, F) \in \mathcal{E}(\mathcal{B})$

- $\mathcal{M} \Vdash [\Gamma]F \Leftrightarrow \mathcal{M}^{+\Gamma} \Vdash F$

- $\mathcal{M} \Vdash [\Gamma^\circ]F \Leftrightarrow \mathcal{M}^{-\Gamma} \Vdash F$

We observe that our shorthand notation is interpreted as expected:

Lemma 3.15. *For a JUP$^\pm$ model $\mathcal{M} = (\mathsf{v}, \mathcal{B})$, and JUP$^\pm$ formulas F, G,*

3.3. Semantics for JUP$^\pm$

- $\mathcal{M} \Vdash F \wedge G \Leftrightarrow (\mathcal{M} \Vdash F$ and $\mathcal{M} \Vdash G)$
- $\mathcal{M} \Vdash F \vee G \Leftrightarrow (\mathcal{M} \Vdash F$ or $\mathcal{M} \Vdash G)$
- $\mathcal{M} \Vdash F \leftrightarrow G \Leftrightarrow (\mathcal{M} \Vdash F$ iff $\mathcal{M} \Vdash G)$

Proof. Straightforward from the definition of truth relation for propositional connectives. □

As with JUP, we need to limit the class of models to obtain soundness and completeness.

Definition 3.16 (Initial model). JUP$^\pm$ model $\mathcal{M} = (\mathsf{v}, \mathcal{B})$ is called *initial* if, for all formulas F, G, $(\mathsf{up}(F), G) \notin \mathcal{B}$.

Definition 3.17 (CS-model). For a constant specification CS, a JUP$^\pm$ model $\mathcal{M} = (\mathsf{v}, \mathcal{B})$ is called a CS-*model* if CS $\subseteq \mathcal{B}$.

Definition 3.18 (Validity). A JUP$^\pm$ formula F is called *valid w.r.t. initial* CS-*models* if, for every initial CS-model $\mathcal{M}$, we have $\mathcal{M} \Vdash F$.
It is denoted as $\Vdash_{\mathsf{JUP}^\pm_{\mathsf{CS}}} F$.

Lemma 3.19 (Properties of updated and rolled-back models). *For any* JUP$^\pm$ *model* $\mathcal{M} = (\mathsf{v}, \mathcal{B})$, *finite sets of formulas* Γ, Δ:

(a) $\mathcal{M}^{+\varnothing} = \mathcal{M}^{-\varnothing} = \mathcal{M}$

(b) $\mathcal{M}^{+\Gamma+\Delta} = \mathcal{M}^{+(\Gamma \cup \Delta)}$

(c) *If* $\mathcal{M}$ *is an initial model*, $\mathcal{M}^{-\Gamma} = \mathcal{M}$

(d) *If* $\mathcal{M}$ *is an initial model*, $\mathcal{M}^{+\Gamma-\Delta} = \mathcal{M}^{+(\Gamma \setminus \Delta)}$

(e) *For any constant specification* CS, *if* $\mathcal{M}$ *is a* CS-*model then so are* $\mathcal{M}^{+\Gamma}, \mathcal{M}^{-\Gamma}$

Proof. (a) $\mathcal{U}_\varnothing = \{(\mathsf{up}(F), F) \mid F \in \varnothing\} = \varnothing$. Therefore,

$$\mathcal{M}^{+\Gamma} = (\mathsf{v}, \mathcal{B}^{+\Gamma}) = (\mathsf{v}, \mathcal{B} \cup \mathcal{U}_\Gamma) = (\mathsf{v}, \mathcal{B} \cup \varnothing) = (\mathsf{v}, \mathcal{B}) = \mathcal{M}$$

$$\mathcal{M}^{-\Gamma} = (\mathsf{v}, \mathcal{B}^{-\Gamma}) = (\mathsf{v}, \mathcal{B} \setminus \mathcal{U}_\Gamma) = (\mathsf{v}, \mathcal{B} \setminus \varnothing) = (\mathsf{v}, \mathcal{B}) = \mathcal{M}$$

3. System JUP$^\pm$

(b) Follows from the following equivalences:
$$\mathcal{M}^{+\Gamma+\Delta} = (\mathsf{v}, \mathcal{B} \cup \mathcal{U}_\Gamma)^{+\Delta}$$
$$= (\mathsf{v}, (\mathcal{B} \cup \mathcal{U}_\Gamma) \cup \mathcal{U}_\Delta)$$
$$= (\mathsf{v}, \mathcal{B} \cup (\mathcal{U}_\Gamma \cup \mathcal{U}_\Delta)) = \mathcal{M}^{+(\Gamma \cup \Delta)}$$

(c) By definition, $\mathcal{M}^{-\Gamma} = (\mathsf{v}, \mathcal{B} \setminus \mathcal{U}_\Gamma)$.

Since $\mathcal{M}$ is assumed to be initial, $\mathcal{B}$ does not contain evidence pairs of the form $(\mathsf{up}(F), F)$ for any F. Therefore, $\mathcal{B} \cap \mathcal{U}_\Gamma = \varnothing$ and $\mathcal{B} \setminus \mathcal{U}_\Gamma = \mathcal{B}$, from which $\mathcal{M}^{-\Gamma} = (\mathsf{v}, \mathcal{B}) = \mathcal{M}$.

(d) By definition,
$$\mathcal{M}^{+\Gamma-\Delta} = (\mathsf{v}, \mathcal{B} \cup \mathcal{U}_\Gamma)^{-\Delta} = (\mathsf{v}, (\mathcal{B} \cup \mathcal{U}_\Gamma) \setminus \mathcal{U}_\Delta)$$

We have $(\mathcal{B} \cup \mathcal{U}_\Gamma) \setminus \mathcal{U}_\Delta = (\mathcal{B} \setminus \mathcal{U}_\Delta) \cup (\mathcal{U}_\Gamma \setminus \mathcal{U}_\Delta)$, so the above can be rewritten as
$$(\mathsf{v}, (\mathcal{B} \cup \mathcal{U}_\Gamma) \setminus \mathcal{U}_\Delta) = (\mathsf{v}, (\mathcal{B} \setminus \mathcal{U}_\Delta) \cup (\mathcal{U}_\Gamma \setminus \mathcal{U}_\Delta))$$
$$= (\mathsf{v}, (\mathcal{B} \setminus \mathcal{U}_\Delta))^{+(\Gamma \setminus \Delta)} = (\mathcal{M}^{-\Delta})^{+(\Gamma \setminus \Delta)}$$

However, since $\mathcal{M}$ is assumed to be initial, we have $\mathcal{M}^{-\Delta} = \mathcal{M}$ from the previous claim. Therefore, $\mathcal{M}^{+\Gamma-\Delta} = \mathcal{M}^{+(\Gamma \setminus \Delta)}$.

(e) Assuming that $\mathcal{M} = (\mathsf{v}, \mathcal{B})$ is a CS-model, or equivalently $\mathsf{CS} \subseteq \mathcal{B}$, we need to show that $\mathsf{CS} \subseteq \mathcal{B}^{+\Gamma}$ and $\mathsf{CS} \subseteq \mathcal{B}^{-\Gamma}$.

The first claim is trivial: if $\mathsf{CS} \subseteq \mathcal{B}$, then $\mathsf{CS} \subseteq \mathcal{B} \cup \mathcal{U}_\Gamma = \mathcal{B}^{+\Gamma}$.

By definition, a constant specification CS cannot contain evidence pairs of the form $(\mathsf{up}(F), F)$ for any F, since the term part must be a constant.

Therefore, $\mathsf{CS} \cap \mathcal{U}_\Gamma = \varnothing$, and from that
$$\mathsf{CS} \cap (\mathcal{B}^{-\Gamma}) = \mathsf{CS} \cap (\mathcal{B} \setminus \mathcal{U}_\Gamma) = (\mathsf{CS} \cap \mathcal{B}) \setminus (\mathsf{CS} \cap \mathcal{U}_\Gamma) = \mathsf{CS} \cap \mathcal{B}$$

Therefore, from $\mathsf{CS} \subseteq \mathcal{B}$ we conclude that $\mathsf{CS} \subseteq \mathcal{B}^{-\Gamma}$.

$\square$

4. Soundness and completeness for $\mathsf{JUP}^{\pm}$

4.1. Soundness for $\mathsf{JUP}^{\pm}$

Theorem 4.1 (Soundness of $\mathsf{JUP}^{\pm}_{\mathsf{CS}}$). *For a given constant specification* CS, *every formula F that is provable in $\mathsf{JUP}^{\pm}_{\mathsf{CS}}$ is valid w.r.t. initial* CS-*models:*

$$\vdash_{\mathsf{JUP}^{\pm}_{\mathsf{CS}}} F \quad \Rightarrow \quad \Vdash_{\mathsf{JUP}^{\pm}_{\mathsf{CS}}} F$$

Proof. Let $\mathcal{M} = (\mathsf{v}, \mathcal{B})$ be an arbitrary initial CS-model. Then we need to show that $\mathcal{M} \Vdash F$.

We assume that F is provable in $\mathsf{JUP}^{\pm}_{\mathsf{CS}}$, therefore has at least one derivation. Proof by induction on the length of derivation of F.

Cases depending on the last step of the proof:

<u>Axiom:</u> The proof is an instance of an axiom. Cases depending on the axiom:

1. (Taut) All instances of propositional tautologies hold under all models, as the propositional part of truth relation is the same as truth-table semantics for classical propositional logic.

2. (App) We need to show that

$$\mathcal{M} \Vdash t : (F \to G) \land s : F \leftrightarrow t \cdot_F s : G$$

By Lemma 3.15, this reduces to proving the following equivalence:

$$(\mathcal{M} \Vdash t : (F \to G) \text{ and } \mathcal{M} \Vdash s : F) \Leftrightarrow \mathcal{M} \Vdash t \cdot_F s : G$$

By definition of the truth relation, we have the following equivalence:

$$(\mathcal{M} \Vdash t : (F \to G) \text{ and } \mathcal{M} \Vdash s : F) \Leftrightarrow \{(t, F \to G), (s, F)\} \subseteq \mathcal{E}(\mathcal{B})$$

4. Soundness and completeness for JUP$^{\pm}$

Finally, by Lemma 3.9,

$$(t \cdot_F s, G) \in \mathcal{E}(\mathcal{B}) \Leftrightarrow \{(t, F \to G), (s, F)\} \subseteq \mathcal{E}(\mathcal{B})$$

Putting the last three equivalences together we obtain the claim.

3. (**Red.1**) We need to show that $\mathcal{M} \Vdash [\Gamma]P \leftrightarrow P$.

   By Lemma 3.15, this reduces to proving the following equivalence:

   $$\mathcal{M} \Vdash [\Gamma]P \Leftrightarrow \mathcal{M} \Vdash P$$

   By definition of the truth relation,

   $$\mathcal{M} \Vdash [\Gamma]P \Leftrightarrow \mathcal{M}^{+\Gamma} \Vdash P$$

   Since $\mathcal{M}^{+\Gamma}$ and $\mathcal{M}$ have the same propositional valuation v, we have:

   $$\mathcal{M}^{+\Gamma} \Vdash P \Leftrightarrow P \in v \Leftrightarrow \mathcal{M} \Vdash P$$

   This shows the required equivalence.

4. (**Red.2**) We need to show that $\mathcal{M} \Vdash [\Gamma]\neg F \leftrightarrow \neg [\Gamma]F$.

   By Lemma 3.15, this reduces to proving the following equivalence:

   $$\mathcal{M} \Vdash [\Gamma]\neg F \Leftrightarrow \mathcal{M} \Vdash \neg[\Gamma]F$$

   By definition of the truth relation,

   $$\mathcal{M} \Vdash [\Gamma]\neg F \Leftrightarrow \mathcal{M}^{+\Gamma} \Vdash \neg F \Leftrightarrow \mathcal{M}^{+\Gamma} \nVdash F$$

   On the other hand,

   $$\mathcal{M} \Vdash \neg[\Gamma]F \Leftrightarrow \mathcal{M} \nVdash [\Gamma]F \Leftrightarrow \mathcal{M}^{+\Gamma} \nVdash F$$

   Together, this shows the required equivalence.

5. (**Red.3**) We need to show that $\mathcal{M} \Vdash [\Gamma](F \to G) \leftrightarrow ([\Gamma]F \to [\Gamma]G)$.

By Lemma 3.15, this reduces to proving the following equivalence:

$$\mathcal{M} \Vdash [\Gamma](F \to G) \Leftrightarrow \mathcal{M} \Vdash [\Gamma]F \to [\Gamma]G$$

By definition of the truth relation,

$$\mathcal{M} \Vdash [\Gamma](F \to G) \Leftrightarrow \mathcal{M}^{+\Gamma} \Vdash F \to G$$
$$\Leftrightarrow (\mathcal{M}^{+\Gamma} \nVdash F \text{ or } \mathcal{M}^{+\Gamma} \Vdash G)$$

On the other hand,

$$\mathcal{M} \Vdash [\Gamma]F \to [\Gamma]G \Leftrightarrow (\mathcal{M} \nVdash [\Gamma]F \text{ or } \mathcal{M} \Vdash [\Gamma]G)$$
$$\Leftrightarrow (\mathcal{M}^{+\Gamma} \nVdash F \text{ or } \mathcal{M}^{+\Gamma} \Vdash G)$$

Together, this shows the required equivalence.

6. (Pers) We need to show that $\mathcal{M} \Vdash t : F \to [\Gamma]t : F$.

   By definitions of the truth relation and the updated model, we have the following equivalences:

   $$\mathcal{M} \Vdash t : F \Leftrightarrow (t, F) \in \mathcal{E}(\mathcal{B})$$

   $$\mathcal{M} \Vdash [\Gamma]t : F \Leftrightarrow \mathcal{M}^{+\Gamma} \Vdash t : F \Leftrightarrow (t, F) \in \mathcal{E}(\mathcal{B} \cup \mathcal{U}_\Gamma)$$

   Since $\mathcal{B} \subseteq \mathcal{B} \cup \mathcal{U}_\Gamma$, by Lemma 3.10, we have

   $$(t, F) \in \mathcal{E}(\mathcal{B}) \Rightarrow (t, F) \in \mathcal{E}(\mathcal{B} \cup \mathcal{U}_\Gamma)$$

   Therefore, we establish that $\mathcal{M} \Vdash t : F \Rightarrow \mathcal{M} \Vdash [\Gamma]t : F$, which, by definition of truth, proves $\mathcal{M} \Vdash t : F \to [\Gamma]t : F$.

7. (Init) We need to show that $\mathcal{M} \Vdash \neg\mathsf{up}(F) : G$.

   By definition of the truth relation,

   $$\mathcal{M} \Vdash \neg\mathsf{up}(F) : G \Leftrightarrow \mathcal{M} \nVdash \mathsf{up}(F) : G \Leftrightarrow (\mathsf{up}(F), G) \notin \mathcal{E}(\mathcal{B})$$

Since up(F) is atomic, by Lemma 3.9

$$(\mathsf{up}(F), G) \notin \mathcal{E}(\mathcal{B}) \Leftrightarrow (\mathsf{up}(F), G) \notin \mathcal{B}$$

The latter is true since $\mathcal{M}$ is assumed to be initial, which shows the claim.

8. (Up) We need to show that $\mathcal{M} \Vdash [\Gamma]\mathsf{up}(F) : F$ assuming $F \in \Gamma$.
By definitions of the truth relation and the updated model,

$$\mathcal{M} \Vdash [\Gamma]\mathsf{up}(F) : F \Leftrightarrow \mathcal{M}^{+\Gamma} \Vdash \mathsf{up}(F) : F \Leftrightarrow (\mathsf{up}(F), F) \in \mathcal{E}(\mathcal{B} \cup \mathcal{U}_\Gamma)$$

By definition of the update set $\mathcal{U}_\Gamma$, we have

$$F \in \Gamma \Rightarrow (\mathsf{up}(F), F) \in \mathcal{U}_\Gamma \subseteq \mathcal{B} \cup \mathcal{U}_\Gamma$$

By Lemma 3.9, $(\mathsf{up}(F), F) \in \mathcal{E}(\mathcal{B} \cup \mathcal{U}_\Gamma)$, which shows the claim.

9. (MC.1) Assume $t \in \mathsf{ATm}$ and either $t \neq \mathsf{up}(F)$ for any F, or $t = \mathsf{up}(F)$ for $F \notin \Gamma$.
We need to show that $\mathcal{M} \Vdash [\Gamma]t : F \to t : F$.
By definition of the truth relation, it is sufficient to show the implication

$$\mathcal{M} \Vdash [\Gamma]t : F \Rightarrow (t, F) \in \mathcal{E}(\mathcal{B})$$

Analogous to the above case,

$$\mathcal{M} \Vdash [\Gamma]t : F \Leftrightarrow (t, F) \in \mathcal{E}(\mathcal{B} \cup \mathcal{U}_\Gamma)$$

By Lemma 3.9, since $t \in \mathsf{ATm}$ we have

$$(t, F) \in \mathcal{E}(\mathcal{B} \cup \mathcal{U}_\Gamma) \Leftrightarrow (t, F) \in \mathcal{B} \cup \mathcal{U}_\Gamma$$

By definition of the update set $\mathcal{U}_\Gamma$,

$$(t, F) \in \mathcal{U}_\Gamma \Leftrightarrow (t = \mathsf{up}(F) \text{ and } F \in \Gamma)$$

Therefore, from our assumption about t follows that $(t, F) \notin \mathcal{U}_\Gamma$.

4.1. Soundness for JUP$^\pm$

Now, assume that $\mathcal{M} \Vdash [\Gamma]t : F$. From the above, it's equivalent to $(t, F) \in \mathcal{B} \cup \mathcal{U}_\Gamma$.

Since $(t, F) \notin \mathcal{U}_\Gamma$, we conclude that $(t, F) \in \mathcal{B}$.

By Lemma 3.9 follows $(t, F) \in \mathcal{E}(\mathcal{B})$, which shows the required implication.

10. (MC.2) We need to show that
$$\mathcal{M} \Vdash [\Gamma]t \cdot_F s : G \leftrightarrow [\Gamma]t : (F \to G) \wedge [\Gamma]s : F$$

By Lemma 3.15 and the definition of truth relation, the claim is equivalent to
$$\mathcal{M}^{+\Gamma} \Vdash t \cdot_F s : G \Leftrightarrow (\mathcal{M}^{+\Gamma} \Vdash t : (F \to G) \text{ and } \mathcal{M}^{+\Gamma} \Vdash s : F)$$

The proof then proceeds as in the case of (App), as it does not depend on the fact that the model is initial.

11. (It) We need to show that $\mathcal{M} \Vdash [\Gamma][\Delta]F \leftrightarrow [\Gamma \cup \Delta]F$.

By Lemma 3.15 and the definition of truth relation, the claim is equivalent to
$$\mathcal{M}^{+\Gamma+\Delta} \Vdash F \Leftrightarrow \mathcal{M}^{+(\Gamma \cup \Delta)} \Vdash F$$

By Lemma 3.19, $\mathcal{M}^{+\Gamma+\Delta} = \mathcal{M}^{+(\Gamma \cup \Delta)}$, which proves the equivalence and the claim.

12. (Roll). We need to show that $\mathcal{M} \Vdash [\Gamma^\circ]F \leftrightarrow F$.

By Lemma 3.15 and the definition of truth relation, this is equivalent to
$$\mathcal{M}^{-\Gamma} \Vdash F \Leftrightarrow \mathcal{M} \Vdash F$$

Since $\mathcal{M}$ is initial, by Lemma 3.19 $\mathcal{M}^{-\Gamma} = \mathcal{M}$, which proves the equivalence and the claim.

13. (Int). We need to show that $\mathcal{M} \Vdash [\Gamma][\Delta^\circ]F \leftrightarrow [\Gamma \setminus \Delta]F$.

By Lemma 3.15 and the definition of truth relation, the claim is equivalent to
$$\mathcal{M}^{+\Gamma-\Delta} \Vdash F \Leftrightarrow \mathcal{M}^{+(\Gamma \setminus \Delta)} \Vdash F$$

Since $\mathcal{M}$ is initial, by Lemma 3.19 $\mathcal{M}^{+\Gamma-\Delta} = \mathcal{M}^{+(\Gamma\setminus\Delta)}$, which proves the equivalence and the claim.

(AN): Assume the proof consists of an application of (AN):

$$\frac{(c,F) \in \mathsf{CS}}{c:F} \text{ (AN)}$$

In that case, we need to show $\mathcal{M} \Vdash c:F$, or equivalently $(c,F) \in \mathcal{E}(\mathcal{B})$. Since $\mathcal{M}$ is a CS-model, applying Lemma 3.9 we obtain

$$(c,F) \in \mathsf{CS} \subseteq \mathcal{B} \subseteq \mathcal{E}(\mathcal{B})$$

This proves the claim.

(MP): Assume the proof terminates in an application of (MP):

$$\frac{F \quad F \to G}{G} \text{ (MP)}$$

By induction hypothesis, we have $\mathcal{M} \Vdash F$ and $\mathcal{M} \Vdash F \to G$, since those subderivations are shorter. By the definition of truth relation, $\mathcal{M} \Vdash F \to G$ is equivalent to $(\mathcal{M} \Vdash F \Rightarrow \mathcal{M} \Vdash G)$. Therefore, we obtain $\mathcal{M} \Vdash G$ and prove the claim. $\square$

4.2. Completeness for JUP$^\pm$

Completeness proof proceeds by using maximal consistent sets for canonical model construction.

Definition 4.2 (Consistency)**.** For a given constant specification CS, an arbitrary set Φ of JUP$^\pm$ formulas is called *consistent* if

$$\mathsf{JUP}^\pm_{\mathsf{CS}} \nvdash \neg(A_1 \wedge \cdots \wedge A_n) \text{ for any finite subset } \{A_1, \ldots, A_n\} \subseteq \Phi$$

A set Φ is called *maximal consistent* if it is consistent, and no proper superset of Φ is.

Remark 4.3 (Lindenbaum Lemma)**.** For every consistent set Φ, there is a maximal consistent set Φ' such that $\Phi \subseteq \Phi'$.

4.2. Completeness for JUP$^\pm$

Lemma 4.4. *Let Φ be a maximal consistent set. Then the following is true:*

- *Φ contains all instances of all axioms.*
- *Φ is closed under* (MP), (AN).
- *$\neg F \in \Phi \Leftrightarrow F \notin \Phi$*
- *$(F \to G) \in \Phi \Leftrightarrow (F \notin \Phi \text{ or } G \in \Phi)$*
- *$(F \leftrightarrow G) \in \Phi \Leftrightarrow (F \in \Phi \Leftrightarrow G \in \Phi)$*

Remark 4.3 and Lemma 4.4 are standard for (maximal) consistent sets; for proofs, see e.g. [KS16].

Given a maximal consistent set, we can create a corresponding model:

Definition 4.5 (Induced JUP$^\pm$ model). Let Φ be a maximal consistent set of formulas. We define its *induced model* as $\mathcal{M}_\Phi := (\mathsf{v}_\Phi, \mathcal{B}_\Phi)$, where

- $\mathsf{v}_\Phi := \Phi \cap \mathsf{Prop}$
- $\mathcal{B}_\Phi := \{(t, F) \mid t : F \in \Phi, t \in \mathsf{ATm}\}$

Lemma 4.6. *For any maximal consistent set of formulas, $\mathcal{M}_\Phi$ is an initial CS-model.*

Proof. By construction, $\mathsf{v}_\Phi \subseteq \mathsf{Prop}$ and $\mathcal{B}_\Phi \subseteq \mathsf{ATm} \times \mathsf{Fm}$, therefore $\mathcal{M}_\Phi$ is a JUP$^\pm$ model.

By Lemma 4.4, Φ is closed under (AN); therefore, for any $(c, F) \in \mathsf{CS}$ we must have $c : F \in \Phi$.

Since $c \in \mathsf{ATm}$, by construction we conclude $(c, F) \in \mathcal{B}_\Phi$, which shows that $\mathcal{M}_\Phi$ is a CS-model.

By Lemma 4.4 every instance of axiom (Init) is in Φ: $\neg\mathsf{up}(F) : G \in \Phi$.
Since Φ is maximal consistent, above is equivalent to $\mathsf{up}(F) : G \notin \Phi$.
$\mathsf{up}(F)$ is atomic, therefore by construction

$$\mathsf{up}(F) : G \notin \Phi \Leftrightarrow (\mathsf{up}(F), G) \notin \mathcal{B}_\Phi$$

Since this holds for arbitrary F and G, this proves that $\mathcal{M}_\Phi$ is initial. $\square$

We need to show that induced models adequately reflect the set they are induced by. The first step in doing so it the canonical evidence lemma:

4. Soundness and completeness for JUP$^\pm$

Lemma 4.7 (Canonical evidence). *For any maximal consistent set Φ,*
$$(t, F) \in \mathcal{E}(\mathcal{B}_\Phi) \quad \Leftrightarrow \quad t : F \in \Phi$$

Proof. Proof by induction on the term t.
  Base: $t \in \mathsf{ATm}$.
  In that case, by construction, $t : F \in \Phi \Leftrightarrow (t, F) \in \mathcal{B}_\Phi$.
  However, since $t \in \mathsf{ATm}$, we can apply Lemma 3.9 to get
$$(t, F) \in \mathcal{B}_\Phi \Leftrightarrow (t, F) \in \mathcal{E}(\mathcal{B}_\Phi)$$

Together, this proves the equivalence for the base case.
Step: $t = r \cdot_G s$ for some terms r, s and formula G.
By Lemma 3.9,
$$(r \cdot_G s, F) \in \mathcal{E}(\mathcal{B}_\Phi) \Leftrightarrow \{(r, G \to F), (s, G)\} \subseteq \mathcal{E}(\mathcal{B}_\Phi)$$

On the other hand, by Lemma 4.4, the following instance of axiom (App) is in Φ:
$$r : (G \to F) \land s : G \leftrightarrow r \cdot_G s : F \in \Phi$$

Using appropriate instances of propositional tautologies that are also in Φ and the fact that Φ is closed under (MP) (again, by Lemma 4.4) we can derive from the previous statement that
$$r \cdot_G s : F \in \Phi \Leftrightarrow \{r : (G \to F), s : G\} \subseteq \Phi$$

Induction hypothesis applies for r and s, and we have
$$(r, G \to F) \in \mathcal{E}(\mathcal{B}_\Phi) \quad \Leftrightarrow \quad r : (G \to F) \in \Phi$$
$$(s, G) \in \mathcal{E}(\mathcal{B}_\Phi) \quad \Leftrightarrow \quad s : G \in \Phi$$

Put together, those equivalences prove the claim for t, concluding the proof. $\square$

In order to proceed with the proof of completeness, we need to define a rank function on functions, which will be used for a proof by induction.

Definition 4.8 (Rank). For a $\mathcal{L}_s^\pm$ term, we inductively define its *term rank* with the following clauses:

4.2. Completeness for JUP$^{\pm}$

- $\mathsf{rk}_{\mathsf{Tm}}(t) := 1$ for $t \in \mathsf{ATm}$.
- $\mathsf{rk}_{\mathsf{Tm}}(t \cdot_F s) := \max(\mathsf{rk}_{\mathsf{Tm}}(t), \mathsf{rk}_{\mathsf{Tm}}(s)) + 1$.

For a $\mathcal{L}_{\mathsf{s}}^{\pm}$ formula, we inductively define its *formula rank* with the following clauses:

- $\mathsf{rk}(P) := 1$ for $P \in \mathsf{Prop}$.
- $\mathsf{rk}(\neg F) := \mathsf{rk}(F) + 1$.
- $\mathsf{rk}(F \to G) := \max(\mathsf{rk}(F), \mathsf{rk}(G)) + 1$.
- $\mathsf{rk}(t : F) := \mathsf{rk}_{\mathsf{Tm}}(t)$.
- $\mathsf{rk}([\Gamma]F) := 2 \cdot \mathsf{rk}(F)$.
- $\mathsf{rk}([\Gamma^{\circ}]F) := 2 \cdot \mathsf{rk}(F)$.

Remark 4.9 (Base rank). The rank functions have the following properties:

- $\mathsf{rk}_{\mathsf{Tm}}(t) \geqslant 1$
- $\mathsf{rk}_{\mathsf{Tm}}(t) = 1 \Rightarrow t \in \mathsf{ATm}$
- $\mathsf{rk}(F) \geqslant 1$
- $\mathsf{rk}(F) = 1 \Rightarrow \begin{cases} F = P & \text{for some } P \in \mathsf{Prop}, \text{ or} \\ F = t : G & \text{for some } t \in \mathsf{ATm}, G \in \mathsf{Fm} \end{cases}$

Lemma 4.10 (Update/rollback reduction). *The formula rank function obeys the following inequalities:*

$$\mathsf{rk}([\Gamma]F) > \mathsf{rk}(F)$$
$$\mathsf{rk}([\Gamma^{\circ}]F) > \mathsf{rk}(F)$$
$$\mathsf{rk}([\Gamma]\neg F) > \mathsf{rk}(\neg[\Gamma]F)$$
$$\mathsf{rk}([\Gamma](F \to G)) > \mathsf{rk}([\Gamma]F \to [\Gamma]G)$$
$$\mathsf{rk}([\Gamma]t \cdot_F s : G) > \mathsf{rk}([\Gamma]t : (F \to G))$$
$$\mathsf{rk}([\Gamma]t \cdot_F s : G) > \mathsf{rk}([\Gamma]s : F)$$
$$\mathsf{rk}([\Gamma][\Delta]F) > \mathsf{rk}([\Gamma \cup \Delta]F)$$
$$\mathsf{rk}([\Gamma][\Delta^{\circ}]F) > \mathsf{rk}([\Gamma \setminus \Delta]F)$$

4. Soundness and completeness for JUP$^\pm$

Proof. By Remark 4.9, $\mathsf{rk}(F) \geqslant 1$ for any formula F, and therefore the first two inequalities hold:

$$\mathsf{rk}([\Gamma]F) = \mathsf{rk}([\Gamma^\circ]F) = 2 \cdot \mathsf{rk}(F) = \mathsf{rk}(F) + \mathsf{rk}(F) \geqslant \mathsf{rk}(F) + 1 > \mathsf{rk}(F)$$

The next 4 ones can be directly observed by computing respective ranks:

$$\begin{aligned}\mathsf{rk}([\Gamma]\neg F) &= 2 \cdot (\mathsf{rk}(F) + 1) \\ &= 2 \cdot \mathsf{rk}(F) + 2 \\ &> 2 \cdot \mathsf{rk}(F) + 1 = \mathsf{rk}(\neg[\Gamma]F)\end{aligned}$$

$$\begin{aligned}\mathsf{rk}([\Gamma](F \to G)) &= 2 \cdot (\max(\mathsf{rk}(F), \mathsf{rk}(G)) + 1) \\ &= \max(2 \cdot \mathsf{rk}(F), 2 \cdot \mathsf{rk}(G)) + 2 \\ &> \max(2 \cdot \mathsf{rk}(F), 2 \cdot \mathsf{rk}(G)) + 1 \\ &= \mathsf{rk}([\Gamma]F \to [\Gamma]G)\end{aligned}$$

$$\begin{aligned}\mathsf{rk}([\Gamma]t \cdot_F s : G) &= 2 \cdot (\max(\mathsf{rk}(t), \mathsf{rk}(s)) + 1) \\ &\geqslant 2 \cdot \mathsf{rk}(t) + 2 \\ &> 2 \cdot \mathsf{rk}(t) \\ &= \mathsf{rk}([\Gamma]t : (F \to G))\end{aligned}$$

$$\begin{aligned}\mathsf{rk}([\Gamma]t \cdot_F s : G) &= 2 \cdot (\max(\mathsf{rk}(t), \mathsf{rk}(s)) + 1) \\ &\geqslant 2 \cdot \mathsf{rk}(s) + 2 \\ &> 2 \cdot \mathsf{rk}(s) \\ &= \mathsf{rk}([\Gamma]s : F)\end{aligned}$$

The final two inequalities again depend on the observation $\mathsf{rk}(F) \geqslant 1$ from Remark 4.9:

$$\begin{aligned}\mathsf{rk}([\Gamma][\Delta]F) &= 4 \cdot \mathsf{rk}(F) \\ &\geqslant 2 \cdot \mathsf{rk}(F) + 2 \\ &> 2 \cdot \mathsf{rk}(F) \\ &= \mathsf{rk}([\Gamma \cup \Delta]F)\end{aligned}$$

4.2. Completeness for JUP$^\pm$

$$\begin{aligned}\mathsf{rk}([\Gamma][\Delta^\circ]F) &= 4 \cdot \mathsf{rk}(F) \\ &\geqslant 2 \cdot \mathsf{rk}(F) + 2 \\ &> 2 \cdot \mathsf{rk}(F) \\ &= \mathsf{rk}([\Gamma \setminus \Delta]F)\end{aligned}$$

$\square$

Finally, we can prove that the induced model fully represents the set of formulas that generated it:

Lemma 4.11 (Truth lemma). *Let Φ be a maximal consistent set of formulas. Then,*

$$F \in \Phi \quad \Leftrightarrow \quad \mathcal{M}_\Phi \Vdash F$$

Proof. Proof proceeds by induction on $\mathsf{rk}(F)$ and case distinction based on the structure of F.

From Remark 4.9, induction base ($\mathsf{rk}(F) = 1$) is covered by subcases $F = P$ and $F = t : G$ for $t \in \mathsf{ATm}$.

- $F = P$ for $P \in \mathsf{Prop}$.

    In this case, by definitions of induced model and truth relation for propositional variables,

    $$P \in \Phi \Leftrightarrow P \in \mathsf{v}_\Phi \Leftrightarrow \mathcal{M}_\Phi \Vdash P$$

- $F = \neg G$.

    By definition of formula rank, observe that $\mathsf{rk}(\neg G) > \mathsf{rk}(G)$ and therefore the induction hypothesis applies for G.

    Since Φ is maximal consistent, by Lemma 4.4 $\neg G \in \Phi \Leftrightarrow G \notin \Phi$.

    Then,

    $$\neg G \in \Phi \Leftrightarrow G \notin \Phi \overset{(\mathsf{IH})}{\Longleftrightarrow} \mathcal{M}_\Phi \nVdash G \Leftrightarrow \mathcal{M}_\Phi \Vdash \neg G$$

- $F = G \to H$.

    By definition of formula rank, observe that $\mathsf{rk}(G \to H) > \mathsf{rk}(G)$ and $\mathsf{rk}(G \to H) > \mathsf{rk}(H)$. Therefore the induction hypothesis applies for both G and H.

4. Soundness and completeness for JUP$^\pm$

Then, using Lemma 4.4,

$$(G \to H) \in \Phi \overset{(4.4)}{\iff} (G \notin \Phi \text{ or } H \in \Phi)$$
$$\overset{(IH)}{\iff} (\mathcal{M}_\Phi \not\Vdash G \text{ or } \mathcal{M}_\Phi \Vdash H)$$
$$\overset{(def)}{\iff} \mathcal{M}_\Phi \Vdash G \to H$$

- $F = t : G$.

  This case is completely covered by Lemma 4.7:

  $$t : G \in \Phi \overset{(4.7)}{\iff} (t, G) \in \mathcal{E}(\mathcal{B}_\Phi) \overset{(def)}{\iff} \mathcal{M}_\Phi \Vdash t : G$$

- $F = [\Gamma]P$.

  This and following cases are making use of Lemma 4.10 in order to apply the induction hypothesis.

  By Lemma 4.4, the following instance of axiom (Red.1) is in Φ:

  $$[\Gamma]P \leftrightarrow P \in \Phi$$

  Applying Lemma 4.4 again, we get

  $$[\Gamma]P \in \Phi \Leftrightarrow P \in \Phi$$

  By Lemma 4.10, $\mathrm{rk}([\Gamma]P) > \mathrm{rk}(P)$, and therefore we can apply the induction hypothesis to P:

  $$[\Gamma]P \in \Phi \Leftrightarrow P \in \Phi \overset{(IH)}{\iff} \mathcal{M}_\Phi \Vdash P$$

  Since $\mathcal{M}_\Phi$ is an initial CS-model, by Theorem 4.1 the same instance of the axiom (Red.1) is true in $\mathcal{M}_\Phi$: $\mathcal{M}_\Phi \Vdash [\Gamma]P \leftrightarrow P$.

  From this and Lemma 3.15 we get the following equivalence:

  $$\mathcal{M}_\Phi \Vdash P \Leftrightarrow \mathcal{M}_\Phi \Vdash [\Gamma]P$$

4.2. Completeness for JUP$^\pm$

Putting equivalences together we obtain the claim.

- $F = [\Gamma]\neg G$.

  By Lemma 4.4, the following instance of axiom (Red.2) is in Φ:

  $$[\Gamma]\neg G \leftrightarrow \neg[\Gamma]G \in \Phi$$

  Applying Lemma 4.4 again, we get

  $$[\Gamma]\neg G \in \Phi \Leftrightarrow \neg[\Gamma]G \in \Phi$$

  By Lemma 4.10, $\mathsf{rk}([\Gamma]\neg G) > \mathsf{rk}(\neg[\Gamma]G)$, and therefore we can apply the induction hypothesis to $\neg[\Gamma]G$:

  $$[\Gamma]\neg G \in \Phi \Leftrightarrow \neg[\Gamma]G \in \Phi \overset{(\mathsf{IH})}{\Longleftrightarrow} \mathcal{M}_\Phi \Vdash \neg[\Gamma]G$$

  As in the previous case, we can use Theorem 4.1 and Lemma 3.15 to obtain the following equivalence from axiom (Red.2):

  $$\mathcal{M}_\Phi \Vdash \neg[\Gamma]G \Leftrightarrow \mathcal{M}_\Phi \Vdash [\Gamma]\neg G$$

  Putting equivalences together we obtain the claim.

- $F = [\Gamma](G \to H)$.

  By Lemma 4.4, the following instance of axiom (Red.3) is in Φ:

  $$[\Gamma](G \to H) \leftrightarrow ([\Gamma]G \to [\Gamma]H) \in \Phi$$

  Applying Lemma 4.4 again, we get

  $$[\Gamma](G \to H) \in \Phi \Leftrightarrow ([\Gamma]G \to [\Gamma]H) \in \Phi$$

  By Lemma 4.10, we get

  $$\mathsf{rk}([\Gamma](G \to H)) > \mathsf{rk}([\Gamma]G \to [\Gamma]G)$$

Therefore we can apply the induction hypothesis to $[\Gamma]G \to [\Gamma]H$:

$$([\Gamma]G \to [\Gamma]H) \in \Phi \overset{(\text{IH})}{\Longleftrightarrow} \mathcal{M}_\Phi \Vdash ([\Gamma]G \to [\Gamma]H)$$

As in the previous case, we can use Theorem 4.1 and Lemma 3.15 to obtain the following equivalence from axiom (**Red.3**):

$$\mathcal{M}_\Phi \Vdash ([\Gamma]G \to [\Gamma]H) \Leftrightarrow \mathcal{M}_\Phi \Vdash [\Gamma](G \to H)$$

Putting equivalences together we obtain the claim.

- $F = [\Gamma]t : G$.

  Sub-cases depending on t:

  - $t = \mathsf{up}(G)$ and $G \in \Gamma$.

    In this case, $F = [\Gamma]\mathsf{up}(G) : G$ is an instance of axiom (**Up**).
    By Lemma 4.4, $[\Gamma]\mathsf{up}(G) : G \in \Phi$.
    By Theorem 4.1, $\mathcal{M}_\Phi \Vdash [\Gamma]\mathsf{up}(G) : G$.
    Since both sides are true, this proves the equivalence in the claim.

  - $t \in \mathsf{ATm}$ but $t \neq \mathsf{up}(G)$ or $G \notin \Gamma$.

    By Lemma 4.4, the following instance of axiom (**Pers**) is in Φ:

    $$t : G \to [\Gamma]t : G \in \Phi$$

    Similarly for the following (**MC.1**) instance:

    $$[\Gamma]t : G \to t : G \in \Phi$$

    Applying Lemma 4.4, we get the following equivalence:

    $$[\Gamma]t : G \in \Phi \Leftrightarrow t : G \in \Phi$$

    By Lemma 4.10, $\mathsf{rk}([\Gamma]t : G) > \mathsf{rk}(t : G)$, and therefore we can apply the induction hypothesis for $t : G$:

    $$t : G \in \Phi \Leftrightarrow \mathcal{M}_\Phi \Vdash t : G$$

4.2. Completeness for JUP$^\pm$

Using Theorem 4.1 for above axiom instances and the definition of truth, we obtain the following equivalence:

$$\mathcal{M}_\Phi \Vdash t : G \Leftrightarrow \mathcal{M}_\Phi \Vdash [\Gamma]t : G$$

Putting equivalences together we obtain the claim.

- $t = r \cdot_H s$.

By Lemma 4.4, the following instance of axiom (MC.2) is in Φ:

$$[\Gamma]r \cdot_H s : G \leftrightarrow [\Gamma]r : (H \to G) \wedge [\Gamma]s : H \in \Phi$$

Again by Lemma 4.4, we obtain the following equivalence:

$$[\Gamma]r \cdot_H s : G \in \Phi \Leftrightarrow ([\Gamma]r : (H \to G) \in \Phi \text{ and } [\Gamma]s : H \in \Phi)$$

By Lemma 4.10, $\mathsf{rk}([\Gamma]r : (H \to G))$ and $\mathsf{rk}([\Gamma]s : H)$ are both strictly less than $\mathsf{rk}([\Gamma]r \cdot_H s : G)$. Therefore, induction hypothesis applies to both, and the equivalence becomes

$$[\Gamma]r \cdot_H s : G \in \Phi \Leftrightarrow (\mathcal{M}_\Phi \Vdash [\Gamma]r : (H \to G) \text{ and } \mathcal{M}_\Phi \Vdash [\Gamma]s : H)$$

By Theorem 4.1 and Lemma 3.15, we obtain the following equivalence from axiom (MC.2):

$$\mathcal{M}_\Phi \Vdash [\Gamma]r \cdot_H s : G$$
$$\Leftrightarrow (\mathcal{M}_\Phi \Vdash [\Gamma]r : (H \to G) \text{ and } \mathcal{M}_\Phi \Vdash [\Gamma]s : H)$$

Putting the equivalences together we obtain the claim.

- $F = [\Gamma][\Delta]G$.

By Lemma 4.4, the following instance of axiom (It) is in Φ:

$$[\Gamma][\Delta]G \leftrightarrow [\Gamma \cup \Delta]G \in \Phi$$

Applying Lemma 4.4, we get the following equivalence:

$$[\Gamma][\Delta]G \in \Phi \Leftrightarrow [\Gamma \cup \Delta]G \in \Phi$$

39

4. Soundness and completeness for JUP$^{\pm}$

By Lemma 4.10, $\text{rk}([\Gamma][\Delta]G) > \text{rk}([\Gamma \cup \Delta]G)$, and therefore we can apply the induction hypothesis to $[\Gamma \cup \Delta]G$:

$$[\Gamma \cup \Delta]G \in \Phi \Leftrightarrow \mathcal{M}_\Phi \Vdash [\Gamma \cup \Delta]G$$

On the other hand, we can use Theorem 4.1 and Lemma 3.15 to obtain the following equivalence from axiom (It):

$$\mathcal{M}_\Phi \Vdash [\Gamma \cup \Delta]G \Leftrightarrow \mathcal{M}_\Phi \Vdash [\Gamma][\Delta]G$$

Putting equivalences together we obtain the claim.

- $F = [\Gamma][\Delta^\circ]G$.

  By Lemma 4.4, the following instance of axiom (Int) is in Φ:

  $$[\Gamma][\Delta^\circ]G \leftrightarrow [\Gamma \setminus \Delta]G \in \Phi$$

  Applying Lemma 4.4, we get the following equivalence:

  $$[\Gamma][\Delta^\circ]G \in \Phi \Leftrightarrow [\Gamma \setminus \Delta]G \in \Phi$$

  By Lemma 4.10, $\text{rk}([\Gamma][\Delta^\circ]G) > \text{rk}([\Gamma \setminus \Delta]G)$, and therefore we can apply the induction hypothesis to $[\Gamma \setminus \Delta]G$:

  $$[\Gamma \setminus \Delta]G \in \Phi \Leftrightarrow \mathcal{M}_\Phi \Vdash [\Gamma \setminus \Delta]G$$

  On the other hand, we can use Theorem 4.1 and Lemma 3.15 to obtain the following equivalence from axiom (Int):

  $$\mathcal{M}_\Phi[\Gamma \setminus \Delta]G \Leftrightarrow \mathcal{M}_\Phi \Vdash [\Gamma][\Delta^\circ]G$$

  Putting equivalences together we obtain the claim.

- $F = [\Gamma^\circ]G$.

  By Lemma 4.4, the following instance of axiom (Roll) is in Φ:

  $$[\Gamma^\circ]G \leftrightarrow G \in \Phi$$

4.2. Completeness for JUP$^\pm$

Applying Lemma 4.4 again, we get

$$[\Gamma^\circ]G \in \Phi \Leftrightarrow G \in \Phi$$

By Lemma 4.10, $\mathsf{rk}([\Gamma^\circ]G) > \mathsf{rk}(G)$, and therefore we can apply the induction hypothesis to G:

$$G \in \Phi \Leftrightarrow \mathcal{M}_\Phi \Vdash G$$

We can use Theorem 4.1 and Lemma 3.15 to obtain the following equivalence from axiom (Roll):

$$\mathcal{M}_\Phi \Vdash G \Leftrightarrow \mathcal{M}_\Phi \Vdash [\Gamma^\circ]G$$

Putting equivalences together we obtain the claim.

□

Equipped with the truth lemma, proving completeness is straightforward:

Theorem 4.12 (Completeness). *Logic* $\mathsf{JUP}^\pm_{\mathsf{CS}}$ *is complete w.r.t. initial* CS-*models, i.e.*

$$\Vdash_{\mathsf{JUP}^\pm_{\mathsf{CS}}} F \quad \Rightarrow \quad \vdash_{\mathsf{JUP}^\pm_{\mathsf{CS}}} F$$

Proof. Proof by contraposition. Assume that $\nvdash_{\mathsf{JUP}^\pm_{\mathsf{CS}}} F$. In that case, $\{\neg F\}$ is a consistent set.

Then, there exists a maximal consistent set Φ containing $\{\neg F\}$, and by Lemma 4.11 $\mathcal{M}_\Phi \Vdash \neg F$, or equivalently $\mathcal{M}_\Phi \nVdash F$.

Therefore, there is a counter-model for F and $\nVdash_{\mathsf{JUP}^\pm_{\mathsf{CS}}} F$. □

5. Rollback properties: history erasing and contraction

5.1. History erasing

As noted in Section 3.1, when chaining updates and rollbacks together, the order in which they are applied matters.

When interpreting a formula of the form $[\Gamma][\Delta^\circ]F$, we apply the "outermost" update with Γ first, then the rollback with Δ, and finally we interpret F in the resulting model.

The following theorem illustrates how a rollback deep in a chain of updates and rollbacks can be interpreted as erasing corresponding functions from all previous operations:

Theorem 5.1 (History Erasing). *Let the symbol σ_i to stand either for "+" or "−" for $i \in \{1, \ldots, n\}$, and define*

$$[\Gamma^{\sigma_i}] := \begin{cases} [\Gamma], & \sigma_i \text{ is "+"} \\ [\Gamma^\circ], & \sigma_i \text{ is "−"} \end{cases}$$

Then, for any constant specification CS, *formula F and finite sets of formulas $\Delta, \Gamma_1, \ldots, \Gamma_n$,*

$$\vdash_{\mathsf{JUP}^\pm_{\mathsf{CS}}} [\Gamma_1^{\sigma_1}] \ldots [\Gamma_n^{\sigma_n}][\Delta^\circ]F \leftrightarrow [(\Gamma_1 \setminus \Delta)^{\sigma_1}] \ldots [(\Gamma_n \setminus \Delta)^{\sigma_n}]F$$

Proof. By Theorem 4.12, it is sufficient to show that the formula is valid w.r.t. initial CS-models:

$$\Vdash_{\mathsf{JUP}^\pm_{\mathsf{CS}}} [\Gamma_1^{\sigma_1}] \ldots [\Gamma_n^{\sigma_n}][\Delta^\circ]F \leftrightarrow [(\Gamma_1 \setminus \Delta)^{\sigma_1}] \ldots [(\Gamma_n \setminus \Delta)^{\sigma_n}]F$$

By definition of validity, it is sufficient to show that for an arbitrary

5. Rollback properties: history erasing and contraction

initial CS-model $\mathcal{M} = (\mathsf{v}, \mathcal{B})$, the formula is true in it:

$$\mathcal{M} \Vdash [\Gamma_1^{\sigma_1}] \ldots [\Gamma_n^{\sigma_n}][\Delta^\circ]F \leftrightarrow [(\Gamma_1 \setminus \Delta)^{\sigma_1}] \ldots [(\Gamma_n \setminus \Delta)^{\sigma_n}]F$$

By Remark 3.15, this reduces to proving the following equivalence:

$$\mathcal{M} \Vdash [\Gamma_1^{\sigma_1}] \ldots [\Gamma_n^{\sigma_n}][\Delta^\circ]F \Leftrightarrow \mathcal{M} \Vdash [(\Gamma_1 \setminus \Delta)^{\sigma_1}] \ldots [(\Gamma_n \setminus \Delta)^{\sigma_n}]F$$

By definition of truth, this becomes

$$\mathcal{M}^{\sigma_1 \Gamma_1 \ldots \sigma_n \Gamma_n - \Delta} \Vdash F \Leftrightarrow \mathcal{M}^{\sigma_1(\Gamma_1 \setminus \Delta) \ldots \sigma_n(\Gamma_n \setminus \Delta)} \Vdash F$$

We will show that $\mathcal{M}^{\sigma_1 \Gamma_1 \ldots \sigma_n \Gamma_n - \Delta} = \mathcal{M}^{\sigma_1(\Gamma_1 \setminus \Delta) \ldots \sigma_n(\Gamma_n \setminus \Delta)}$, from which the previous equivalence trivially follows.

For σ either "+" or "−" and bases $\mathcal{B}_1, \mathcal{B}_2$, define

$$\mathcal{B}_1 \, \sigma \, \mathcal{B}_2 := \begin{cases} \mathcal{B}_1 \cup \mathcal{B}_2, & \sigma_i \text{ is "+"} \\ \mathcal{B}_1 \setminus \mathcal{B}_2, & \sigma_i \text{ is "−"} \end{cases}$$

We will assume this to be left-associative: $\mathcal{B}_1 \, \sigma \, \mathcal{B}_2 \, \sigma' \, \mathcal{B}_3 = (\mathcal{B}_1 \, \sigma \, \mathcal{B}_2) \, \sigma' \, \mathcal{B}_3$. Then, by definition of model updates and rollbacks, we have

$$\mathcal{M}^{\sigma_1 \Gamma_1 \ldots \sigma_n \Gamma_n - \Delta} = (\mathsf{v}, (\mathcal{B} \, \sigma_1 \, \mathcal{U}_{\Gamma_1} \ldots \sigma_n \, \mathcal{U}_{\Gamma_n}) \setminus \mathcal{U}_\Delta)$$
$$\mathcal{M}^{\sigma_1(\Gamma_1 \setminus \Delta) \ldots \sigma_n(\Gamma_n \setminus \Delta)} = (\mathsf{v}, \mathcal{B} \, \sigma_1 \, \mathcal{U}_{(\Gamma_1 \setminus \Delta)} \ldots \sigma_n \, \mathcal{U}_{(\Gamma_n \setminus \Delta)})$$

Therefore, it is sufficient to show that

$$(\mathcal{B} \, \sigma_1 \, \mathcal{U}_{\Gamma_1} \ldots \sigma_n \, \mathcal{U}_{\Gamma_n}) \setminus \mathcal{U}_\Delta = \mathcal{B} \, \sigma_1 \, \mathcal{U}_{(\Gamma_1 \setminus \Delta)} \ldots \sigma_n \, \mathcal{U}_{(\Gamma_n \setminus \Delta)}$$

Induction on n.

<u>Base</u>: $n = 0$. We need to show $\mathcal{B} \setminus \mathcal{U}_\Delta = \mathcal{B}$.

Since $\mathcal{M}$ was assumed to be initial, $\mathcal{B}$ does not contain evidence pairs of the form $(\mathsf{up}(G), G)$; on the other hand, $\mathcal{U}_\Delta$ contains only evidence pairs of this form. Therefore, nothing is removed and $\mathcal{B} \setminus \mathcal{U}_\Delta = \mathcal{B}$.

<u>Step</u>: We assume the statement is true for n by induction hypothesis,

and prove the statement for $n+1$:

$$(\mathcal{B}\,\sigma_1\,\mathcal{U}_{\Gamma_1}\ldots\sigma_n\,\mathcal{U}_{\Gamma_n}\,\sigma_{n+1}\,\mathcal{U}_{\Gamma_{n+1}})\setminus\mathcal{U}_\Delta$$
$$=\mathcal{B}\,\sigma_1\,\mathcal{U}_{(\Gamma_1\setminus\Delta)}\ldots\sigma_n\,\mathcal{U}_{(\Gamma_n\setminus\Delta)}\,\sigma_{n+1}\,\mathcal{U}_{(\Gamma_{n+1}\setminus\Delta)}$$

Cases based on σ_{n+1}:

(a) σ_{n+1} is "+". In this case,

$$(\mathcal{B}\,\sigma_1\,\mathcal{U}_{\Gamma_1}\ldots\sigma_n\,\mathcal{U}_{\Gamma_n}\,\sigma_{n+1}\,\mathcal{U}_{\Gamma_{n+1}})\setminus\mathcal{U}_\Delta$$
$$=((\mathcal{B}\,\sigma_1\,\mathcal{U}_{\Gamma_1}\ldots\sigma_n\,\mathcal{U}_{\Gamma_n})\cup\mathcal{U}_{\Gamma_{n+1}})\setminus\mathcal{U}_\Delta$$
$$=((\mathcal{B}\,\sigma_1\,\mathcal{U}_{\Gamma_1}\ldots\sigma_n\,\mathcal{U}_{\Gamma_n})\setminus\mathcal{U}_\Delta)\cup(\mathcal{U}_{\Gamma_{n+1}}\setminus\mathcal{U}_\Delta)$$
$$\stackrel{(\mathrm{IH})}{=}(\mathcal{B}\,\sigma_1\,\mathcal{U}_{(\Gamma_1\setminus\Delta)}\ldots\sigma_n\,\mathcal{U}_{(\Gamma_n\setminus\Delta)})\cup(\mathcal{U}_{\Gamma_{n+1}}\setminus\mathcal{U}_\Delta)$$

Observe that

$$\mathcal{U}_{\Gamma_{n+1}}\setminus\mathcal{U}_\Delta=\{(\mathsf{up}(G),G)\mid G\in\Gamma_{n+1}\}\setminus\{(\mathsf{up}(G),G)\mid G\in\Delta\}$$
$$=\{(\mathsf{up}(G),G)\mid G\in(\Gamma_{n+1}\setminus\Delta)\}$$
$$=\mathcal{U}_{(\Gamma_{n+1}\setminus\Delta)}$$

Therefore,

$$(\mathcal{B}\,\sigma_1\,\mathcal{U}_{(\Gamma_1\setminus\Delta)}\ldots\sigma_n\,\mathcal{U}_{(\Gamma_n\setminus\Delta)})\cup(\mathcal{U}_{\Gamma_{n+1}}\setminus\mathcal{U}_\Delta)$$
$$=(\mathcal{B}\,\sigma_1\,\mathcal{U}_{(\Gamma_1\setminus\Delta)}\ldots\sigma_n\,\mathcal{U}_{(\Gamma_n\setminus\Delta)})\cup\mathcal{U}_{(\Gamma_{n+1}\setminus\Delta)}$$
$$=\mathcal{B}\,\sigma_1\,\mathcal{U}_{(\Gamma_1\setminus\Delta)}\ldots\sigma_n\,\mathcal{U}_{(\Gamma_n\setminus\Delta)}\,\sigma_{n+1}\,\mathcal{U}_{(\Gamma_{n+1}\setminus\Delta)}$$

Putting those equalities together shows the claim.

(b) σ_{n+1} is "-". In this case,

$$(\mathcal{B}\,\sigma_1\,\mathcal{U}_{\Gamma_1}\ldots\sigma_n\,\mathcal{U}_{\Gamma_n}\,\sigma_{n+1}\,\mathcal{U}_{\Gamma_{n+1}})\setminus\mathcal{U}_\Delta$$
$$=((\mathcal{B}\,\sigma_1\,\mathcal{U}_{\Gamma_1}\ldots\sigma_n\,\mathcal{U}_{\Gamma_n})\setminus\mathcal{U}_{\Gamma_{n+1}})\setminus\mathcal{U}_\Delta$$
$$=((\mathcal{B}\,\sigma_1\,\mathcal{U}_{\Gamma_1}\ldots\sigma_n\,\mathcal{U}_{\Gamma_n})\setminus\mathcal{U}_\Delta)\setminus(\mathcal{U}_{\Gamma_{n+1}}\setminus\mathcal{U}_\Delta)$$
$$\stackrel{(\mathrm{IH})}{=}(\mathcal{B}\,\sigma_1\,\mathcal{U}_{(\Gamma_1\setminus\Delta)}\ldots\sigma_n\,\mathcal{U}_{(\Gamma_n\setminus\Delta)})\setminus(\mathcal{U}_{\Gamma_{n+1}}\setminus\mathcal{U}_\Delta)$$

5. Rollback properties: history erasing and contraction

As above, we have $\mathcal{U}_{\Gamma_{n+1}} \setminus \mathcal{U}_\Delta = \mathcal{U}_{(\Gamma_{n+1} \setminus \Delta)}$

Therefore,

$$(\mathcal{B}\, \sigma_1\, \mathcal{U}_{(\Gamma_1 \setminus \Delta)} \ldots \sigma_n\, \mathcal{U}_{(\Gamma_n \setminus \Delta)}) \setminus (\mathcal{U}_{\Gamma_{n+1}} \setminus \mathcal{U}_\Delta)$$
$$= (\mathcal{B}\, \sigma_1\, \mathcal{U}_{(\Gamma_1 \setminus \Delta)} \ldots \sigma_n\, \mathcal{U}_{(\Gamma_n \setminus \Delta)}) \setminus \mathcal{U}_{(\Gamma_{n+1} \setminus \Delta)}$$
$$= \mathcal{B}\, \sigma_1\, \mathcal{U}_{(\Gamma_1 \setminus \Delta)} \ldots \sigma_n\, \mathcal{U}_{(\Gamma_n \setminus \Delta)}\, \sigma_{n+1}\, \mathcal{U}_{(\Gamma_{n+1} \setminus \Delta)}$$

Putting those equalities together shows the claim.

$\square$

5.2. Rollback and contraction

In [KS13], the update operator for JUP is shown to satisfy AGM postulates for belief expansion [Gär88], and this naturally extends to the update operator for JUP$^\pm$.

This raises the question whether the rollback operation, which removes beliefs, can be seen as belief contraction. To discuss this, we need to introduce the notion of (induced) belief sets (following definitions from [KS13]):

Definition 5.2 (Belief set). A set of formulas $X \subseteq \mathsf{Fm}$ is called a *belief set* if it is closed under (MP):

$$\{F, F \to G\} \subseteq X \Rightarrow G \in X$$

Definition 5.3 (Induced belief set). For a JUP$^\pm$ model $\mathcal{M}$, define its *induced belief set* $\square_\mathcal{M}$ as

$$\square_\mathcal{M} := \{F \mid \mathcal{M} \Vdash t : F \text{ for some } t \in \mathsf{Tm}\}$$

Lemma 5.4 (Induced belief set). *For any* JUP$^\pm$ *model* $\mathcal{M} = (\mathsf{v}, \mathcal{B})$, $\square_\mathcal{M}$ *is a belief set.*

Proof. Same proof as in [KS13]:
 Assume $\{F, F \to G\} \subseteq \square_\mathcal{M}$.
 By definition, this means that $\mathcal{M} \Vdash t : F$ and $\mathcal{M} \Vdash s : (F \to G)$ for some $t, s \in \mathsf{Tm}$, or equivalently $\{(s, F \to G), (t, F)\} \subseteq \mathcal{E}(\mathcal{B})$.

5.2. Rollback and contraction

By Lemma 3.9, we can infer $(s \cdot_F t, G) \in \mathcal{E}(\mathcal{B})$, which is equivalent to $\mathcal{M} \Vdash s \cdot_F t : G$.
By definition of induced belief set, we conclude $G \in \square_\mathcal{M}$. □

Lemma 5.5 (Belief monotonicity). *For any two models $\mathcal{M} = (\mathsf{v}, \mathcal{B})$ and $\mathcal{M}' = (\mathsf{v}', \mathcal{B}')$, the following holds:*

$$\mathcal{B} \subseteq \mathcal{B}' \quad \Rightarrow \quad \square_\mathcal{M} \subseteq \square_{\mathcal{M}'}$$

Proof. Assume $\mathcal{B} \subseteq \mathcal{B}'$ and let $F \in \square_\mathcal{M}$.
By definition of induced beliefs, there exists a term t such that $\mathcal{M} \Vdash t : F$.
By definition of the truth relation, this is equivalent to $(t, F) \in \mathcal{E}(\mathcal{B})$.
By Lemma 3.10, $\mathcal{E}(\mathcal{B}) \subseteq \mathcal{E}(\mathcal{B}')$ and therefore $(t, F) \in \mathcal{E}(\mathcal{B}')$ or equivalently $\mathcal{M}' \Vdash t : F$.
With this, we conclude $F \in \square_{\mathcal{M}'}$, which shows the claim. □

By applying an update operation, we extend induced belief sets, as we introduce new reasons for beliefs. First, we define an operator on induced belief sets:

Definition 5.6 (Expansion). For a model $\mathcal{M}$ and a formula F, define the *expansion* of the belief set $\square_\mathcal{M}$ by F as

$$\square_\mathcal{M} \oplus F := \square_{\mathcal{M}+F}$$

Then, [KS13] shows the following lemma:

Lemma 5.7 (Belief expansion). *$\square_\mathcal{M} \oplus F$ is the smallest belief set that contains both F and all of $\square_\mathcal{M}$.*

It's natural to ask whether the rollback operation, that removes reasons for beliefs, acts like belief contraction in AGM sense [AGM85]. Let's define a "naive" contraction operator:

Definition 5.8 (Naive contraction). For a model $\mathcal{M}$ and a formula F, define the *naive contraction* of the belief set $\square_\mathcal{M}$ by F as

$$\square_\mathcal{M} \ominus_1 F := \square_{\mathcal{M}-F}$$

Notably, naive contraction in conjunction with expansion has the *recovery* property:

5. Rollback properties: history erasing and contraction

Lemma 5.9 (Recovery for $\ominus_1$). *For any model $\mathcal{M} = (\mathsf{v}, \mathcal{B})$ and any formula F,*
$$\Box_\mathcal{M} \subseteq (\Box_\mathcal{M} \ominus_1 F) \oplus F$$

Proof. By definition, $(\Box_\mathcal{M} \ominus_1 F) \oplus F = \Box_{\mathcal{M}^{-F+F}}$

Note that $\mathcal{M}^{-F+F} = (\mathsf{v}, \mathcal{B} \setminus \mathcal{U}_{\{F\}} \cup \mathcal{U}_{\{F\}}) = (\mathsf{v}, \mathcal{B} \cup \mathcal{U}_{\{F\}})$.

Since $\mathcal{B} \subseteq \mathcal{B} \cup \mathcal{U}_{\{F\}}$, by Lemma 5.5 $\Box_\mathcal{M} \subseteq \Box_{\mathcal{M}^{-F+F}}$, which shows the claim. $\square$

However, one of the requirements for classic belief contraction is *success*: contracting with a formula should remove it from a belief set. AGM postulates accept that this is impossible in general, as tautologies are in a sense "built-in" and cannot be successfully contracted.

However, even for non-tautologies rollback may not succeed to remove a belief. Take an arbitrary model $\mathcal{M}$ and consider the following update with 3 formulas, where G is not a tautology (and even stronger, $G \notin \Box_\mathcal{M}$).

$$G \in \Box_{\mathcal{M}+\{G, F, F \to G\}}$$

If we now try to apply naive contraction with G, it will not remove it from the belief set, since it's closed under (**MP**):

$$G \in \Box_{\mathcal{M}+\{G, F, F \to G\}} \ominus_1 G = \Box_{\mathcal{M}+\{F, F \to G\}}$$

The reason for this difference is that, for any formula H, both $\Box_\mathcal{M} \oplus H$ and $\Box_\mathcal{M} \ominus_1 H$ add or remove one particular reason (namely, $\mathsf{up}(H)$) to believe in a formula H. For a formula to appear in the induced belief set, any single reason is enough, which is why updates always succeed.

Since justification logic can provide multiple independent reasons to believe in a formula, removing just one one may fail to completely remove the belief, as evidenced by the above example.

However, we can roll back more updates to successfully remove G from the belief set:

$$G \notin \Box_{\mathcal{M}+\{G, F, F \to G\} - \{G, F, F \to G\}} = \Box_\mathcal{M}$$

This naturally raises the question which formulas can, for a given model, be removed from its induced belief set by some rollback operation. For a certain natural class of models, we can give a precise answer.

5.2. Rollback and contraction

Definition 5.10 (Initial part). For a model $\mathcal{M} = (\mathsf{v}, \mathcal{B})$, define its *initial part* $\mathcal{M}^\circ := (\mathsf{v}, \mathcal{B}^\circ)$, where

$$\mathcal{B}^\circ := \{(t, F) \mid (t, F) \in \mathcal{B}, t \neq \mathsf{up}(G) \text{ for any } G \in \mathsf{Fm}\}$$

Essentially, taking the initial part filters out any update terms from the resulting evidence relation. Directly from the definition of initial models we get the following remark:

Remark 5.11 (Initial part). For any model $\mathcal{M}$, its initial part $\mathcal{M}^\circ$ is an initial model.

Definition 5.12 (Reachable model). We call a model $\mathcal{M}$ *reachable* if there exists a finite set $\Gamma \subseteq \mathsf{Fm}$ such that $\mathcal{M} = (\mathcal{M}^\circ)^{+\Gamma}$.

Reachable models are a natural class of models: by Lemma 3.19 and Remark 5.11 those are precisely the models obtainable from initial models by finite sequences of update and rollback operations.

For such models, we can show precisely what formulas can be removed from induced belief sets:

Theorem 5.13 (Retractability). *For a reachable model $\mathcal{M} = (\mathsf{v}, \mathcal{B})$ and a formula F, the following holds:*

(a) *If $F \in \square_{\mathcal{M}^\circ}$, then for any finite $\Delta \subseteq \mathsf{Fm}$, $F \in \square_{\mathcal{M}^{-\Delta}}$.*

(b) *If $F \notin \square_{\mathcal{M}^\circ}$, then there exists a finite $\Delta \subseteq \mathsf{Fm}$ such that $F \notin \square_{\mathcal{M}^{-\Delta}}$.*

Proof. $\mathcal{M}$ is reachable, so there exists a finite set Γ such that $\mathcal{M} = (\mathcal{M}^\circ)^{+\Gamma}$.

(a) For an arbitrary Δ, by Remark 5.11 and Lemma 3.19 we have

$$\mathcal{M}^{-\Delta} = (\mathcal{M}^\circ)^{+\Gamma-\Delta} = (\mathcal{M}^\circ)^{+(\Gamma\setminus\Delta)} = (\mathsf{v}, \mathcal{B}^\circ \cup \mathcal{U}_{(\Gamma\setminus\Delta)})$$

Since $\mathcal{B}^\circ \subseteq \mathcal{B}^\circ \cup \mathcal{U}_{(\Gamma\setminus\Delta)}$, by Lemma 5.5 we conclude $\square_{\mathcal{M}^\circ} \subseteq \square_{\mathcal{M}^{-\Delta}}$. If we assume that $F \in \square_{\mathcal{M}^\circ}$, this shows $F \in \square_{\mathcal{M}^{-\Delta}}$.

(b) Take $\Delta := \Gamma$.

Then, by Lemma 3.19,

$$\mathcal{M}^{-\Delta} = (\mathcal{M}^\circ)^{+\Gamma-\Gamma} = (\mathcal{M}^\circ)^{+(\Gamma\setminus\Gamma)} = (\mathcal{M}^\circ)^{+\varnothing} = \mathcal{M}^\circ$$

5. Rollback properties: history erasing and contraction

Therefore, by assumption, $F \notin \mathcal{M}^\circ = \mathcal{M}^{-\Delta}$.

$\square$

We have demonstrated that, for reachable models, there is a set of beliefs that are potentially contractible by rollbacks: those not included in the beliefs induced by initial part. We shall call them *acquired beliefs*:

Definition 5.14 (Acquired beliefs). For a model $\mathcal{M}$, define its set of *acquired beliefs* as

$$\mathcal{A}(\mathcal{M}) := \Box_\mathcal{M} \setminus \Box_{\mathcal{M}^\circ}$$

We can try and define a belief contraction operator based on rollbacks that meets the requirements of *shielded contraction* [FH01], i.e. contraction limited to a subset of all formulas.

Definition 5.15 (Full contraction). For a model $\mathcal{M}$ and a formula F, define the *full contraction* of the belief set $\Box_\mathcal{M}$ by F as

$$\Box_\mathcal{M} \ominus_2 F := \begin{cases} \Box_{\mathcal{M}^\circ}, & F \in \mathcal{A}(\mathcal{M}) \\ \Box_\mathcal{M}, & F \notin \mathcal{A}(\mathcal{M}) \end{cases}$$

Note that for a reachable model $\mathcal{M} = (\mathcal{M}^\circ)^{+\Gamma}$, we have $\mathcal{M}^\circ = \mathcal{M}^{-\Gamma}$, and therefore full contraction can be expressed in terms of model rollback.

We can show that full contraction satisfies the conditions of *relative success* and *persistence* with $\mathcal{A}(\mathcal{M})$ being the set of retractable sentences:

Lemma 5.16 (Relative success and persistence for $\ominus_2$). *For any model $\mathcal{M}$ and formulas F, G, the following holds:*

(a) *If $F \in \mathcal{A}(\mathcal{M})$, then $F \notin \Box_\mathcal{M} \ominus_2 F$.*

(b) *If $F \notin \mathcal{A}(\mathcal{M})$, then $\Box_\mathcal{M} = \Box_\mathcal{M} \ominus_2 F$.*

(c) *If $F \in \Box_\mathcal{M} \ominus_2 F$, then $F \in \Box_\mathcal{M} \ominus_2 G$.*

Proof. (a) If $F \in \mathcal{A}(\mathcal{M})$, then $\Box_\mathcal{M} \ominus_2 F = \Box_{\mathcal{M}^\circ}$.

If we assume $F \in \Box_\mathcal{M} \ominus_2 F$, we get $F \in \Box_{\mathcal{M}^\circ}$.

This contradicts $F \in \mathcal{A}(\mathcal{M}) = \Box_\mathcal{M} \setminus \Box_{\mathcal{M}^\circ}$, which shows the claim by contradiction.

(b) If $F \notin \mathcal{A}(\mathcal{M})$, then $\Box_\mathcal{M} \ominus_2 F = \Box_\mathcal{M}$ by definition.

(c) Assume $F \in \Box_\mathcal{M} \ominus_2 F$.
From the above, $F \notin \mathcal{A}(\mathcal{M}) = \Box_\mathcal{M} \setminus \Box_{\mathcal{M}^\circ}$.
However since $F \notin \mathcal{A}(\mathcal{M})$, we have $\Box_\mathcal{M} \ominus_2 F = \Box_\mathcal{M}$ by definition of $\ominus_2$, and therefore $F \in \Box_\mathcal{M}$.
From $F \notin \Box_\mathcal{M} \setminus \Box_{\mathcal{M}^\circ}$ and $F \in \Box_\mathcal{M}$ we conclude that $F \in \Box_{\mathcal{M}^\circ}$.
Since $\Box_\mathcal{M} \ominus_2 G$ is either $\Box_\mathcal{M}$ or $\Box_{\mathcal{M}^\circ}$, in either case $F \in \Box_\mathcal{M} \ominus_2 G$ which shows the claim.

$\Box$

However, the full contraction operator removes too many beliefs, as it fails the test for *recovery*: in general $\Box_\mathcal{M} \not\subseteq (\Box_\mathcal{M} \ominus_2 F) \oplus F$.

We can introduce a more generic form of rollback-based contraction operator:

Definition 5.17 (*f*-contraction). Let f be a function that maps a model and a formula to a finite set of formulas.

Then, for a model $\mathcal{M}$ and a formula F, define the *f-contraction* of the belief set $\Box_\mathcal{M}$ by F as

$$\Box_\mathcal{M} \ominus_f F := \begin{cases} \Box_{\mathcal{M} - f(\mathcal{M}, F)}, & F \in \mathcal{A}(\mathcal{M}) \\ \Box_\mathcal{M}, & F \notin \mathcal{A}(\mathcal{M}) \end{cases}$$

$\ominus_1$ and, for a reachable model $\mathcal{M} = (\mathcal{M}^\circ)^{+\Gamma}$, $\ominus_2$ can be expressed in terms of f-contraction:

- $f_1(\mathcal{M}, F) := \{F\}$ (since $\Box_{\mathcal{M} - F} = \Box_\mathcal{M}$ for $F \notin \mathcal{A}(\mathcal{M})$)

- $f_2(\mathcal{M}, F) := \Gamma$

We can show that arbitrary f-contraction fits some of the requirements of a shielded contraction operator:

Lemma 5.18 (*f*-contraction properties). *For a model* $\mathcal{M} = (\mathsf{v}, \mathcal{B})$ *and arbitrary formulas* F, G, *any* f-*contraction operator has the following properties:*

(a) *(Closure)* $\Box_\mathcal{M} \ominus_f F$ *is a belief set.*

(b) *(Inclusion)* $\Box_\mathcal{M} \ominus_f F \subseteq \Box_\mathcal{M}$.

(c) *(Vacuity)* If $F \notin \Box_\mathcal{M}$, then $\Box_\mathcal{M} \subseteq \Box_\mathcal{M} \ominus_f F$.

Proof.

(a) *(Closure)* $\Box_\mathcal{M} \ominus_f F$ is a belief set as an induced belief set of either $\mathcal{M}^{-f(\mathcal{M},F)}$ or $\mathcal{M}$.

(b) *(Inclusion)* We need to show that $\Box_\mathcal{M} \ominus_f F \subseteq \Box_\mathcal{M}$.

If $F \notin \mathcal{A}(\mathcal{M})$, then by definition $\Box_\mathcal{M} \ominus_f F = \Box_\mathcal{M} \subseteq \Box_\mathcal{M}$.

If $F \in \mathcal{A}(\mathcal{M})$, then $\Box_\mathcal{M} \ominus_f F = \Box_{\mathcal{M}^{-f(\mathcal{M},F)}}$.

Note that $\mathcal{M}^{-f(\mathcal{M},F)} = (\mathsf{v}, \mathcal{B} \setminus \mathcal{U}_{f(\mathcal{M},F)})$.

Since $\mathcal{B} \setminus \mathcal{U}_{f(\mathcal{M},F)} \subseteq \mathcal{B}$, by Lemma 5.5 $\Box_{\mathcal{M}^{-f(\mathcal{M},F)}} \subseteq \Box_\mathcal{M}$, which shows the claim.

(c) *(Vacuity)* Assume $F \notin \Box_\mathcal{M}$.

Therefore, $F \notin \mathcal{A}(\mathcal{M}) \subseteq \Box_\mathcal{M}$, and by definition $\Box_\mathcal{M} \ominus_f F = \Box_\mathcal{M}$, which shows the claim.

$\Box$

This shows that, in general, f-contractions are good candidates for a shielded contraction operator.

We should remark that in the study of belief revision, it's usually assumed that belief sets are closed with respect to more expressive logic - normally classic propositional logic. As in [KS13], with a suitably expressive constant specification, belief sets of **CS**-models have such properties.

First, we define what we mean by suitably expressive constant specification:

Definition 5.19 (Appropriate constant specification). A constant specification **CS** is called

- *propositionally appropriate* if, for every propositional tautology F, there exists a constant c such that $(c, F) \in \mathsf{CS}$;

- *axiomatically appropriate* if, for every instance F of $\mathsf{JUP}^\pm$ axioms, there exists a constant c such that $(c, F) \in \mathsf{CS}$;

5.2. Rollback and contraction

- $\mathsf{JUP}^{\pm}_{\mathsf{CS}}$-*appropriate* if is is axiomatically appropriate and, for every $(c, F) \in \mathsf{CS}$, there exists a constant c' such that $(c', c : F) \in \mathsf{CS}$.

Propositionally appropriate specification means that every classical propositional axiom instance is in a CS-model's belief set, and combined with being closed under (MP) as per Lemma 5.4 this gives the full strength of propositional reasoning. Similarly, axiomatically appropriate constant specifications give all instances of $\mathsf{JUP}^{\pm}$ axioms, and $\mathsf{JUP}^{\pm}_{\mathsf{CS}}$-appropriateness closes with respect to (CS). Formally:

Definition 5.20. A set of formulas X is *closed with respect to reasoning in logic Th* if, for any formula F that can be derived using axioms and rules of Th as well as formulas from X (notation $X \vdash_{Th} F$), $F \in X$.

Lemma 5.21. *For a CS-model $\mathcal{M}$, formula F and arbitrary f-contraction operator $\ominus_f$, the following holds:*

(a) *If CS is propositionally appropriate, then belief sets $\Box_{\mathcal{M}}$, $\Box_{\mathcal{M}} \oplus F$, and $\Box_{\mathcal{M}} \ominus_f F$ are closed with respect to reasoning in classic propositional logic (i.e.* (Prop) + (MP)*).*

(b) *If CS is axiomatically appropriate, then belief sets $\Box_{\mathcal{M}}$, $\Box_{\mathcal{M}} \oplus F$, and $\Box_{\mathcal{M}} \ominus_f F$ are closed with respect to reasoning in $\mathsf{JUP}^{\pm}_{\varnothing}$ (i.e. $\mathsf{JUP}^{\pm}_{\mathsf{CS}}$ with an empty constant specification).*

(c) *If CS is $\mathsf{JUP}^{\pm}_{\mathsf{CS}}$-appropriate, then belief sets $\Box_{\mathcal{M}}$, $\Box_{\mathcal{M}} \oplus F$, $\Box_{\mathcal{M}} \ominus_f F$ are closed with respect to reasoning in $\mathsf{JUP}^{\pm}_{\mathsf{CS}}$.*

Proof. Since $\mathcal{M}$ is a CS-model, by Lemma 3.19 so are $\mathcal{M}^{-f(\mathcal{M},F)}$ and $\mathcal{M}^{+F}$.

Both $\Box_{\mathcal{M}} \oplus F$ and $\Box_{\mathcal{M}} \ominus_f F$ are, by definition, belief sets of either $\mathcal{M}^{+F}$, $\mathcal{M}$ or $\mathcal{M}^{-f(\mathcal{M},F)}$, which we've just shown to be CS-models.

Therefore, if we can show the claim for an arbitrary CS-model $\mathcal{M}$, it would also cover cases for $\Box_{\mathcal{M}} \oplus F$ and $\Box_{\mathcal{M}} \ominus_f F$.

(a) Let CS be a propositionally appropriate constant specification.

Take an arbitrary formula G for which there exists a derivation in classic propositional logic using $\Box_{\mathcal{M}}$ as a theory.

We will show that $G \in \Box_{\mathcal{M}}$ by induction on the length of derivation of G, which shows that $\Box_{\mathcal{M}}$ is closed with respect to reasoning in classic propositional logic.

53

5. Rollback properties: history erasing and contraction

Base: The last step of the derivation is a formula from $\Box_\mathcal{M}$.

This means that $G \in \Box_\mathcal{M}$.

Axiom: The last step of the derivation is an axiom instance of classical propositional logic (in other words, an instance of (Prop)).

This means that G is a propositional tautology.

By propositional appropriateness of CS, there exists a constant c such that $(c, G) \in$ CS.

Since $\mathcal{M}$ is a CS-model, $(c, G) \in \mathcal{B}$ and therefore $\mathcal{M} \Vdash c : G$. By definition, this implies $G \in \Box_\mathcal{M}$.

Step: The last step of the derivation is an instance of (MP) for some formula H:

$$\frac{H \quad H \to G}{G} \text{ (MP)}$$

Since sub-derivations of H and $H \to G$ are shorter, by induction hypothesis we conclude that $\{H, H \to G\} \subseteq \Box_\mathcal{M}$.

By Lemma 5.4, $\Box_\mathcal{M}$ is closed under (MP) and therefore we conclude $G \in \Box_\mathcal{M}$.

(b) Let CS be an axiomatically appropriate constant specification.

Take an arbitrary formula G for which there exists a derivation in $\mathsf{JUP}^\pm_\varnothing$ using $\Box_\mathcal{M}$ as a theory.

We will show that $G \in \Box_\mathcal{M}$ by induction on the length of derivation of G, which shows that $\Box_\mathcal{M}$ is closed with respect to reasoning in $\mathsf{JUP}^\pm_\varnothing$.

Base: The last step of the derivation is a formula from $\Box_\mathcal{M}$.

This means that $G \in \Box_\mathcal{M}$.

Axiom: The last step of the derivation is an axiom instance of $\mathsf{JUP}^\pm$.

This means that G is an axiom instance of $\mathsf{JUP}^\pm$.

By axiomatic appropriateness of CS, there exists a constant c such that $(c, G) \in$ CS.

Since $\mathcal{M}$ is a CS-model, $(c, G) \in \mathcal{B}$ and therefore $\mathcal{M} \Vdash c : G$. By definition, this implies $G \in \Box_\mathcal{M}$.

Step: The last step of the derivation is an instance of either (MP) or (AN).

The sub-case for (MP) is analogous to the previous case.

For (AN), notice that there are no valid instances of (AN) for $\mathsf{JUP}^{\pm}_{\varnothing}$, since the premise $(c, H) \in \varnothing$ cannot be fulfilled. Therefore, this sub-case is impossible.

(c) Let CS be a $\mathsf{JUP}^{\pm}_{\mathsf{CS}}$-appropriate constant specification.

Take an arbitrary formula G for which there exists a derivation in $\mathsf{JUP}^{\pm}_{\mathsf{CS}}$ using $\Box_\mathcal{M}$ as a theory.

We will show that $G \in \Box_\mathcal{M}$ by induction on the length of derivation of G, which shows that $\Box_\mathcal{M}$ is closed with respect to reasoning in $\mathsf{JUP}^{\pm}_{\mathsf{CS}}$.

The proof is completely analogous to the last case except for the sub-case for (AN).

Step, (AN): The last step of the derivation is an instance of (AN):

$$\frac{(c, H) \in \mathsf{CS}}{c : H} \text{ (AN)}$$

In this case, $G = c : H$ and $(c, H) \in \mathsf{CS}$. By $\mathsf{JUP}^{\pm}_{\mathsf{CS}}$-appropriateness of CS, there exists a constant c' such that $(c', c : H) \in \mathsf{CS}$. As shown above, this entails $c : H \in \Box_\mathcal{M}$, which completes the proof.

$\square$

$\ominus_1$ and $\ominus_2$ are, in a way, lower and upper limits of possible contractions. $\ominus_1$ satisfies the recovery postulate but its set of retractable sentences is in general smaller than $\mathcal{A}(\mathcal{M})$. $\ominus_2$ guarantees persistence and relative success for the whole set $\mathcal{A}(\mathcal{M})$, but does not satisfy the recovery postulate.

It is plausible that there exists a function f "between" them that would satisfy, at the same time, recovery, persistence and relative success requirements for the full set of retractable sentences $\mathcal{A}(\mathcal{M})$, possibly by requiring a sufficiently expressive constant specification to ensure that belief sets are closed with respect to some required strength of reasoning.

It's important to note that both recovery and success postulates are not universally accepted as requirements for belief contraction; shielded

5. Rollback properties: history erasing and contraction

contraction is an attempt to provide alternative to the success postulate, and there may be alternatives to the recovery postulate that are more suitable for $\mathsf{JUP}^{\pm}$.

The existence of such optimal f and whether it would fulfill other postulates for shielded contraction from [FH01] are open questions. They warrant further study of belief dynamics of updates and rollbacks in $\mathsf{JUP}^{\pm}$.

Part III.

Nominals

6. From updates to nominals

As explained in the introduction, one of the goals of the research that forms the basis of this thesis was to obtain an axiomatization of dynamic justification logic with updates that does not use the subscript in its term application.

The goal of this part is to present the system of justification logic with nominals JN_V and prove soundness and completeness results with respect to two different semantics. However, we would like to first give some informal insight into the design decisions behind JN_V as intended simplified version of JUP.

By dropping the subscript, JUP loses the ability to formulate axioms (App) and (MC.2) as equivalences. If we try to naively modify them to be one-way implications, the system loses the ability to reduce justification formulas with application terms to justification formulas with subterms.

This, in turn, means that generated model semantics must be changed to allow bases containing non-atomic justification terms. In this case, the original completeness proof fails, as it relied on the ability to apply (MC.2) backwards.

In order to better understand how update terms behave in the absence of subscript, it was decided to approach the problem from another angle. Instead of a dynamic justification logic, we will consider a "snapshot" of a state after several updates, with special terms called *nominals* playing the role of the update terms.

For a nominal to emulate the update term, we need to recall what update terms justify. After an update with a set Γ that contains a formula F, the corresponding update term $\mathsf{up}(F)$ starts being accepted as justification for the formula F:

$$\vdash [\Gamma]\mathsf{up}(F) : F$$

On the other hand, for any other formula $G \neq F$, the update term

6. From updates to nominals

provably doesn't justify it:

$$\vdash \neg[\Gamma]\mathsf{up}(F) : G$$

Therefore, to emulate update terms after an update with Γ, we need the nominal n_F for $F \in \Gamma$ to justify its respective formula and provably justify nothing else:

$$\vdash n_F : F$$
$$\vdash \neg n_F : G, \quad G \neq F$$

While having atomic models would be ideal, at this point the subscript-free application suggests generated models with non-atomic bases.

However, this approach needs careful tuning of allowed term structure. If arbitrary applications are allowed in the basis, we may end up with anomalous justifications that cannot be satisfied in any atomic models:

$$n_P \cdot t : F$$

This can be satisfied in a non-atomic model, but to satisfy it in an atomic model n_P needs to justify $G \to F$ for some G, which is impossible since $P \neq G \to F$ for any G.

Therefore, we want to exclude such anomalous justifications from our system. We would need many axioms to cover all possible anomalies for arbitrary nominals; as a result, it was decided to simplify the situation we are expecting to model even further and restrict nominals (and therefore, the updates we emulate) to propositional variables: namely, we will have nominals n_P for P in some finite set $\mathsf{V} \subseteq \mathsf{Prop}$.

In case of propositional nominals, the above application is the only case that can not be decomposed in any way into subterms that still yields the original justification. We will call a term *normal* if it contains no subterms of the form $n_P \cdot t$, and will explicitly forbid such terms in both the axiom system and semantics.

Keeping with the goal of eventually obtaining semantics with atomic bases, consider the following justification:

$$t \cdot n_P : F$$

Since n_P can only justify P, in order for the above justification to be obtained from an atomic basis, t must specifically justify $P \to F$. This requirement will have to be added to the axiom system and reflected in the semantics.

At this point, the axiom system and corresponding non-atomic generated model semantics could be shown to be sound and complete. However, in trying to obtain atomic models, another anomaly was noticed. Consider the following formula:

$$x \cdot (y \cdot n_P) : Q \land z : P \land \neg (x \cdot (y \cdot z) : Q)$$

It is satisfiable in a non-atomic model. However, assume we want to find an atomic model that satisfies it.

In this case, $x : (X \to Q)$ for some formula X, and thus $y : (P \to X)$ as we decompose the evidence term from first conjunct. However, we must also have $z : P$ from the second conjunct. In this case, we can use z in place of n_P and infer $x \cdot (y \cdot z) : Q$, which means we cannot satisfy this formula (see Figure 1).

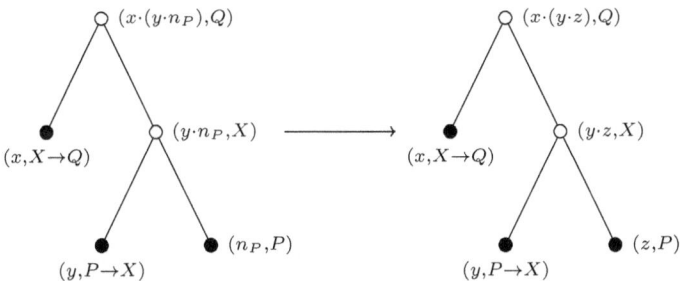

Figure 1: Non-satisfiability example

In order to overcome this anomaly, which has to do with substituting terms for nominals, we introduced explicit substitutions into the language and required, both on the axiom level and in the semantics, that every substitution of terms for nominals must preserve truth. Informally:

$$\vdash t : F \land s : P \to t_{[s \to n_P]} : F$$

61

6. From updates to nominals

With the axiom system containing enough safeguards against anomalous applications, we were able to indirectly show completeness with respect to atomic models, on condition of *locally finite* constant specification.

7. System $\mathsf{JN_V}$

7.1. Language $\mathcal{L}_n^V$

Start with countably many *propositional variables* ($\mathsf{Prop}_{\mathcal{L}_n^V}$): $\{P, \ldots\}$, and fix a finite subset $\mathsf{V} \subseteq \mathsf{Prop}_{\mathcal{L}_n^V}$.

Definition 7.1 (Terms). Define the set of *terms* $\mathsf{Tm}_{\mathcal{L}_n^V}$ inductively:

- (Countably many) term variables: $x, \ldots$
- (Countably many) term constants: $c, \ldots$
- Nominals for propositional variables: n_P for $P \in \mathsf{V}$
- Application: $t \cdot s$ for $t, s \in \mathsf{Tm}_{\mathcal{L}_n^V}$

Atomic terms $\mathsf{ATm}_{\mathcal{L}_n^V}$ are terms without application: variables, constants and nominals.

As before, term application is considered left-associative, and we will drop the subscript $\mathcal{L}_n^V$ for the rest of Part III.

Definition 7.2 (Subterms). Define $\mathrm{sub}(t)$ to be the set of all *subterms* of $t \in \mathsf{Tm}$ in the usual sense:

- $\mathrm{sub}(t) := \{t\}$ for $t \in \mathsf{ATm}$
- $\mathrm{sub}(t \cdot s) := \mathrm{sub}(t) \cup \mathrm{sub}(s) \cup \{t \cdot s\}$

Definition 7.3 (Formulas). Define the set of *formulas* $\mathsf{Fm}_{\mathcal{L}_n^V}$ inductively:

- Propositional variables: $P \in \mathsf{Prop}$
- Implication: $F \to G$ for $F, G \in \mathsf{Fm}_{\mathcal{L}_n^V}$
- Negation: $\neg F$ for $F \in \mathsf{Fm}_{\mathcal{L}_n^V}$

7. System JN$_V$

- Evidence: $t : F$ for $t \in \mathsf{Tm}_{\mathcal{L}_n^V}, F \in \mathsf{Fm}_{\mathcal{L}_n^V}$

We will need to explicitly express substitutions of terms for nominals to formulate one of the axioms and the semantics. For this, we introduce a notion of *basic substitution*, representing a single such substitution:

Definition 7.4 (Basic substitutions). A triple $\langle i, P, s \rangle$, where $i \in \mathbb{N}$, $P \in \mathsf{V}$ and $s \in \mathsf{Tm}$, is called a *basic substitution*.

We denote the set of all basic substitutions as $\mathcal{S}$, and will use the syntactic variable σ to denote them.

Informally, $\langle i, P, s \rangle t$ denotes the result of substituting s for the $(i+1)$-th occurrence of n_P in t.

Formally, define the number of occurrences $\text{num}_P(t)$ of n_P in t:

- $\text{num}_P(n_P) = 1$,
- $\text{num}_P(t) = 0$, $t \in \mathsf{ATm} \setminus \{n_P\}$,
- $\text{num}_P(t \cdot s) = \text{num}_P(t) + \text{num}_P(s)$

Finally, define $\langle i, P, s \rangle t$:

- $\langle i, P, s \rangle n_P = \begin{cases} s & i = 0 \\ n_P & i > 0 \end{cases}$

- $\langle i, P, s \rangle t = t$, $t \in \mathsf{ATm} \setminus \{n_P\}$,

- $\langle i, P, s \rangle (t \cdot q) = \begin{cases} (\langle i, P, s \rangle t) \cdot q & i < \text{num}_P(t) \\ t \cdot (\langle i - \text{num}_P(t), P, s \rangle q) & i \geqslant \text{num}_P(t) \end{cases}$

As a shorthand, define $\langle i, P, s \rangle {\downarrow} t := \langle i - \text{num}_P(t), P, s \rangle$ for $i \geqslant \text{num}_P(t)$.

Lemma 7.5 (Non-applicable substitutions). *For any basic substitution $\sigma = \langle i, P, s \rangle$ and term t such that $i \geqslant \text{num}_P(t)$, we have $\sigma t = t$.*

Proof. Proof by induction on t for arbitrary $\sigma = \langle i, P, s \rangle$ with $i \geqslant \text{num}_P(t)$.
  Base: $t \in \mathsf{ATm}$.
  There are 2 sub-cases:

(a) $t = n_P$. In this case, we have $\text{num}_P(t) = 1$ and therefore by assumption we conclude $i \geqslant 1 > 0$.
    Therefore, $\sigma t = n_P = t$.

(b) $t \in \mathsf{ATm} \setminus \{n_P\}$. In that case, the claim follows immediately.

Step: $t = q \cdot r$.
We have $i \geqslant \mathrm{num}_P(t) = \mathrm{num}_P(q) + \mathrm{num}_P(r) \geqslant \mathrm{num}_P(q)$.
Therefore,

$$\sigma t = \langle i, P, s \rangle (q \cdot r) = q \cdot (\langle i - \mathrm{num}_P(q), P, s \rangle r)$$

Observe that $i - \mathrm{num}_P(q) \geqslant \mathrm{num}_P(q) + \mathrm{num}_P(r) - \mathrm{num}_P(q) = \mathrm{num}_P(r)$.
Therefore, induction hypothesis applies to r and $\langle i - \mathrm{num}_P(q), P, s \rangle$:

$$\langle i - \mathrm{num}_P(q), P, s \rangle r = r$$

From this we conclude $\sigma t = q \cdot r = t$. $\square$

Definition 7.6 (Term complexity). We define the *complexity* of a term $\mathrm{cmp}(t)$ as the number of applications:

$$\mathrm{cmp}(t) = 0 \text{ for } t \in \mathsf{ATm}$$

$$\mathrm{cmp}(t \cdot s) = \mathrm{cmp}(t) + \mathrm{cmp}(s) + 1$$

Lemma 7.7. *Term complexity has the following properties:*

(a) $t \in \mathsf{ATm} \Leftrightarrow \mathrm{cmp}(t) = 0$

(b) For any term t, $\mathrm{cmp}(\sigma t) \geqslant \mathrm{cmp}(t)$ for any basic substitution σ

Proof.

(a) Trivial from definition of cmp.

(b) Induction on t. Let $\sigma = \langle i, P, s \rangle$.
  - Base: $t \in \mathsf{ATm}$. Then $\mathrm{cmp}(\sigma t) \geqslant 0 = \mathrm{cmp}(t)$
  - Step: $t = r \cdot q$. Then

  $$\sigma t = \begin{cases} (\sigma r) \cdot q & i < \mathrm{num}_P(r) \\ r \cdot ((\sigma \!\downarrow\! r) q) & i \geqslant \mathrm{num}_P(r) \end{cases}$$

7. System JN_V

Therefore,

$$\mathrm{cmp}(\sigma t) = \begin{cases} \mathrm{cmp}((\sigma r) \cdot q) = \mathrm{cmp}(\sigma r) + \mathrm{cmp}(q) + 1 \\ \qquad i < \mathrm{num}_P(r), \\ \mathrm{cmp}(r \cdot ((\sigma \downarrow r)q)) = \mathrm{cmp}(r) + \mathrm{cmp}((\sigma \downarrow r)q) + 1 \\ \qquad i \geqslant \mathrm{num}_P(r) \end{cases}$$

By induction hypothesis,

$$\mathrm{cmp}(\sigma r) \geqslant \mathrm{cmp}(r) \text{ and } \mathrm{cmp}((\sigma \downarrow r)q) \geqslant \mathrm{cmp}(q)$$

Therefore, in either case we have

$$\mathrm{cmp}(\sigma t) \geqslant \mathrm{cmp}(r) + \mathrm{cmp}(q) + 1 = \mathrm{cmp}(r \cdot q)$$

$\square$

7.2. Logic JN_V

Definition 7.8. The logic of JN_V is a Hilbert-style deduction system defined by the following sets of axioms and rules ($P \in \mathsf{V}$ in all schemes):

Axioms:

1. All classical propositional tautologies   (Prop)
2. $t : (F \to G) \wedge s : F \to t \cdot s : G$ (App)
3. $n_P : P$ (N_+)
4. $\neg n_P : F$ for $F \neq P$ (N_−)
5. $\neg t : F$ for all t with $n_P \cdot s \in \mathrm{sub}(t)$ (N_◁)
6. $t \cdot n_P : F \to t : (P \to F)$ (N_▷)
7. $t : F \wedge s : P \to \sigma t : F$ where $\sigma = \langle i, P, s \rangle$ (for all i)   (N_σ)

Rule:

$$\frac{F \quad F \to G}{G} \text{ (MP)}$$

7.3. Semantics for $\mathsf{JN_V}$

Definition 7.9. A *constant specification* CS is a set of evidence pairs of the form $(c, c_1 : \ldots : c_n : F)$ (including $n = 0$, i.e. (c, F)), where F is an axiom instance of $\mathsf{JN_V}$.

Definition 7.10. For a given constant specification CS, the logic of $\mathsf{JN_{CS}^V}$ is defined by the axioms and rules of $\mathsf{JN_V}$ plus the rule:

$$\frac{(c, F) \in \mathsf{CS}}{c : F} \text{ (AN)}$$

Definition 7.11. A formula F is *provable in* $\mathsf{JN_{CS}^V}$ (notation: $\vdash_{\mathsf{JN_{CS}^V}} F$) iff it can be derived from axioms and rules of $\mathsf{JN_{CS}^V}$.

7.3. Semantics for $\mathsf{JN_V}$

Definition 7.12 (Basis). A *basis* is an arbitrary set of evidence pairs $\mathcal{B} \subseteq \mathsf{Tm} \times \mathsf{Fm}$.

Definition 7.13 (Subterm set). For a basis $\mathcal{B}$, define its *subterm set*, $\mathrm{sub}(\mathcal{B})$, as

$$\mathrm{sub}(\mathcal{B}) := \bigcup_{(t,F) \in \mathcal{B}} \mathrm{sub}(t)$$

Definition 7.14 (Evidence closure). Take a basis $\mathcal{B} \subseteq \mathsf{Tm} \times \mathsf{Fm}$. For a set $X \subseteq \mathsf{Tm} \times \mathsf{Fm}$ define an operator $\mathrm{cl}_\mathcal{B}^{\mathsf{JN_V}}(X)$ inductively by:

- $(t, F) \in \mathcal{B} \Rightarrow (t, F) \in \mathrm{cl}_\mathcal{B}^{\mathsf{JN_V}}(X)$ (equivalently, $\mathcal{B} \subseteq \mathrm{cl}_\mathcal{B}^{\mathsf{JN_V}}(X)$)

- $(t, F \to G) \in X, (s, F) \in X \Rightarrow (t \cdot s, G) \in \mathrm{cl}_\mathcal{B}^{\mathsf{JN_V}}(X)$

- $(t, F) \in X, (s, P) \in X \Rightarrow (\langle i, P, s \rangle t, F) \in \mathrm{cl}_\mathcal{B}^{\mathsf{JN_V}}(X)$ for $i \in \mathbb{N}, P \in \mathsf{V}$

Note that $\mathrm{cl}_\mathcal{B}^{\mathsf{JN_V}}$ is a monotone operator on $\mathcal{P}(\mathsf{Tm} \times \mathsf{Fm})$ and therefore has a least fixpoint by Knaster–Tarski theorem [Tar55].

Definition 7.15 (Evidence relation). For a $\mathcal{B} \subseteq \mathsf{Tm} \times \mathsf{Fm}$, define the (minimal) evidence relation $\mathcal{E}^{\mathsf{JN_V}}(\mathcal{B})$ as the l.f.p. of $\mathrm{cl}_\mathcal{B}$.

We will drop the superscript $\mathsf{JN_V}$ from $\mathrm{cl}_\mathcal{B}^{\mathsf{JN_V}}$ and $\mathcal{E}^{\mathsf{JN_V}}$ for the rest of the Part III.

7. System JN$_V$

Lemma 7.16. *Evidence relation $\mathcal{E}(\mathcal{B})$ has the following properties:*

(a) $\mathcal{B} \subseteq \mathcal{E}(\mathcal{B})$.

(b) $\{(t, F \to G), (s, F)\} \subseteq \mathcal{E}(\mathcal{B}) \Rightarrow (t \cdot s, G) \in \mathcal{E}(\mathcal{B})$.

(c) $\{(t, F), (s, P)\} \subseteq \mathcal{E}(\mathcal{B}) \Rightarrow (\langle i, P, s \rangle t, F) \in \mathcal{E}(\mathcal{B})$ for all $i \in \mathbb{N}, P \in \mathsf{V}$.

Proof. $\mathcal{E}(\mathcal{B})$ is a fixpoint of $\mathsf{cl}_\mathcal{B}$.
In particular, $\mathsf{cl}_\mathcal{B}(\mathcal{E}(\mathcal{B})) \subseteq \mathcal{E}(\mathcal{B})$.
From this and the definition of $\mathsf{cl}_\mathcal{B}$ we directly obtain the claims. □

In the rest of the proofs, we will frequently appeal to the following induction principle:

Lemma 7.17 (Induction on buildup of $\mathcal{E}(\mathcal{B})$). *Let P be a predicate on $\mathsf{Tm} \times \mathsf{Fm}$. Assume it has the following properties:*

- *(Base) For all $(t, F) \in \mathcal{B}$, $\mathsf{P}(t, F)$ is true.*

- *(Application) For all $t, s \in \mathsf{Tm}$ and $F, G \in \mathsf{Fm}$,*

$$\mathsf{P}(t, F \to G) \text{ and } \mathsf{P}(s, F) \Rightarrow \mathsf{P}(t \cdot s, G)$$

- *(Substitution) For all $t, s \in \mathsf{Tm}$, $F \in \mathsf{Fm}$ and $P \in \mathsf{V}$,*

$$\mathsf{P}(t, F) \text{ and } \mathsf{P}(s, P) \Rightarrow \mathsf{P}(\langle i, P, s \rangle t, F)$$

Then, for all $(t, F) \in \mathcal{E}(\mathcal{B})$, $\mathsf{P}(t, F)$ is true.

Proof. Let $\mathcal{E}_0 := \varnothing$, $\mathcal{E}_{n+1} := \mathsf{cl}_\mathcal{B}(\mathcal{E}_n)$.
Since $\mathsf{cl}_\mathcal{B}$ is monotone, and $\mathcal{E}_0 = \varnothing \subseteq \mathcal{E}_1$, it follows that $\mathcal{E}_n \subseteq \mathcal{E}_{n+1}$ for all $n \in \mathbb{N}$.
We will show that for all $n \in \mathbb{N}$, $\mathsf{P}(t, F)$ is true for all $(t, F) \in \mathcal{E}_n$.
Induction on n:
Base: $n = 0$. In this case, the claim is vacuously true, as $\mathcal{E}_0 = \varnothing$.
Step: We need to show that $\mathsf{P}(t, F)$ is true for all $(t, F) \in \mathcal{E}_{n+1}$, and induction hypothesis applies for n.
Assume $(t, F) \in \mathcal{E}_{n+1} = \mathsf{cl}_\mathcal{B}(\mathcal{E}_n)$. By definition of $\mathsf{cl}_\mathcal{B}$, there are three cases:

- $(t, F) \in \mathcal{B}$. In that case, $\mathsf{P}(t, F)$ is true by property (Base).
- $(t, F) = (s \cdot r, G)$, with $(s, H \to G) \in \mathcal{E}_n, (r, H) \in \mathcal{E}_n$.
  By induction hypothesis, $\mathsf{P}(t, H \to G)$ and $\mathsf{P}(s, H)$ are true.
  Therefore, claim follows by property (Application).
- $(t, F) = (\langle i, P, s \rangle r, G)$, with $(r, G) \in \mathcal{E}_n, (s, P) \in \mathcal{E}_n$.
  By induction hypothesis, $\mathsf{P}(r, G)$ and $\mathsf{P}(s, P)$ are true.
  Therefore, claim follows by property (Substitution).

Define $\mathcal{E}_\omega := \bigcup_{n \in \mathbb{N}} \mathcal{E}_n$.
For any $(t, F) \in \mathcal{E}_\omega$, there exists an n such that $(t, F) \in \mathcal{E}_n$.
Therefore, for all $(t, F) \in \mathcal{E}_\omega$, $\mathsf{P}(t, F)$ is true.
We can show that $\mathcal{E}_\omega$ is a fixpoint of $\mathsf{cl}_\mathcal{B}$:

- $\mathcal{E}_\omega \subseteq \mathsf{cl}_\mathcal{B}(\mathcal{E}_\omega)$.
  Assume $(t, F) \in \mathcal{E}_\omega$.
  Then there exists an n such that $(t, F) \in \mathcal{E}_n$. Moreover, $n > 0$ since $\mathcal{E}_0 = \varnothing$.
  Therefore, $(t, F) \in \mathcal{E}_n = \mathsf{cl}_\mathcal{B}(\mathcal{E}_{n-1}) \subseteq \mathsf{cl}_\mathcal{B}(\mathcal{E}_\omega)$ by monotonicity of $\mathsf{cl}_\mathcal{B}$.

- $\mathsf{cl}_\mathcal{B}(\mathcal{E}_\omega) \subseteq \mathcal{E}_\omega$.
  Assume $(t, F) \in \mathsf{cl}_\mathcal{B}(\mathcal{E}_\omega)$. By definition of $\mathsf{cl}_\mathcal{B}$, there are three cases:
  - $(t, F) \in \mathcal{B}$. In that case, $(t, F) \in \mathsf{cl}_\mathcal{B}(\varnothing) = \mathcal{E}_1 \subseteq \mathcal{E}_\omega$.
  - $(t, F) = (s \cdot r, G)$, with $(s, H \to G) \in \mathcal{E}_\omega, (r, H) \in \mathcal{E}_\omega$.
    There exist n, m such that $(s, H \to G) \in \mathcal{E}_n, (r, H) \in \mathcal{E}_m$.
    From this, $(s, H \to G) \in \mathcal{E}_{\max(n,m)}, (r, H) \in \mathcal{E}_{\max(n,m)}$, since $\{\mathcal{E}_n\}$ is an increasing sequence of sets.
    Therefore,
    $$(t, F) = (s \cdot r, G) \in \mathsf{cl}_\mathcal{B}(\mathcal{E}_{\max(n,m)}) = \mathcal{E}_{\max(n,m)+1} \subseteq \mathcal{E}_\omega$$
  - $(t, F) = (\langle i, P, s \rangle r, G)$, with $(r, G) \in \mathcal{E}_\omega, (s, P) \in \mathcal{E}_\omega$.
    There exist n, m such that $(r, G) \in \mathcal{E}_n, (s, P) \in \mathcal{E}_m$.

From this, $(r, G) \in \mathcal{E}_{\max(n,m)}, (s, P) \in \mathcal{E}_{\max(n,m)}$, since $\{\mathcal{E}_n\}$ is an increasing sequence of sets.

Therefore,
$$(t, F) = (\langle i, P, s\rangle r, G) \in \mathsf{cl}_\mathcal{B}(\mathcal{E}_{\max(n,m)}) = \mathcal{E}_{\max(n,m)+1} \subseteq \mathcal{E}_\omega$$

Since $\mathcal{E}(\mathcal{B})$ is the l.f.p. of $\mathsf{cl}_\mathcal{B}$, we get $\mathcal{E}(\mathcal{B}) \subseteq \mathcal{E}_\omega$. Therefore, for any $(t, F) \in \mathcal{E}(\mathcal{B})$, we have $(t, F) \in \mathcal{E}_\omega$ and subsequently $\mathsf{P}(t, F)$ is true. □

Definition 7.18 (Propositional valuation). A *propositional valuation* is an arbitrary subset of propositional variables $\mathsf{v} \subseteq \mathsf{Prop}$.

Definition 7.19 (Models). A *model* is a pair $\mathcal{M} = (\mathsf{v}, \mathcal{B})$ with propositional valuation $\mathsf{v} \subseteq \mathsf{Prop}$ and basis $\mathcal{B} \subseteq \mathsf{Tm} \times \mathsf{Fm}$.

Definition 7.20 (Truth). For a model $\mathcal{M} = (\mathsf{v}, \mathcal{B})$, and a formula F, the relation $\mathcal{M} \Vdash F$ is defined inductively by:

- $\mathcal{M} \Vdash P \Leftrightarrow P \in \mathsf{v}$
- $\mathcal{M} \Vdash \neg F \Leftrightarrow \mathcal{M} \nVdash F$
- $\mathcal{M} \Vdash F \to G \Leftrightarrow (\mathcal{M} \nVdash F \text{ or } \mathcal{M} \Vdash G)$
- $\mathcal{M} \Vdash t : A \Leftrightarrow (t, A) \in \mathcal{E}(\mathcal{B})$

We observe that our shorthand notation is interpreted as expected:

Lemma 7.21. *For a* $\mathsf{JN_V}$ *model* $\mathcal{M} = (\mathsf{v}, \mathcal{B})$, *and formulas* F, G,

- $\mathcal{M} \Vdash F \wedge G \Leftrightarrow (\mathcal{M} \Vdash F \text{ and } \mathcal{M} \Vdash G)$
- $\mathcal{M} \Vdash F \vee G \Leftrightarrow (\mathcal{M} \Vdash F \text{ or } \mathcal{M} \Vdash G)$
- $\mathcal{M} \Vdash F \leftrightarrow G \Leftrightarrow (\mathcal{M} \Vdash F \text{ iff } \mathcal{M} \Vdash G)$

Proof. Straightforward from the definition of truth relation for propositional connectives. □

Additional restrictions on the model are required for the semantics to be sound w.r.t. the logic JN.

7.3. Semantics for JN$_V$

Definition 7.22. We call a term t *normal* if $(n_P \cdot s) \notin \text{sub}(t)$ for any $P \in \mathsf{V}, s \in \mathsf{Tm}$.

Consider a set of restrictions:

(R.1) $(n_P, F) \in \mathcal{B} \Leftrightarrow F = P$ (for $P \in \mathsf{V}$)

(R.2) $n_P \cdot s \notin \text{sub}(\mathcal{B})$, or equivalently t is normal for all $(t, F) \in \mathcal{B}$.

(R.3) $(t \cdot n_P, F) \notin \mathcal{B}$

Note that the restriction (R.3) allows $t \cdot n_P$ to occur deeper in the term, i.e. $((t \cdot n_P) \cdot s, F)$ is a valid evidence pair for (R.3).

Definition 7.23 (Intermediate models). We call models with restrictions (R.1)—(R.3) on $\mathcal{B}$ *intermediate*.
A formula F is called *valid w.r.t. intermediate models* (notation: $\Vdash_i F$) iff $\mathcal{M} \Vdash F$ for all intermediate models $\mathcal{M}$.

Definition 7.24 (Atomic models). Let $\mathcal{B}_a \subseteq \mathsf{ATm} \times \mathsf{Fm}$ that satisfies (R.1).
We call a model of the form $\mathcal{M}_a := (\mathsf{v}, \mathcal{B}_a)$ *atomic*.
A formula F is called *valid w.r.t. atomic models* (notation: $\Vdash_a F$) iff $\mathcal{M} \Vdash F$ for all atomic models $\mathcal{M}$.

Lemma 7.25. *Any atomic model is also an intermediate model.*

Proof. Let $\mathcal{M}_a := (\mathsf{v}, \mathcal{B}_a)$ be an atomic model.
We need to show that restrictions (R.1)—(R.3) hold.

(R.1) By definition of an atomic model.

(R.2) $\mathcal{B}_a \subseteq \mathsf{ATm} \times \mathsf{Fm}$, therefore for all $(t, F) \in \mathcal{B}_a$ we have

$$\text{sub}(t) = \{t\} \subseteq \mathsf{ATm}$$

Therefore, $n_P \cdot s \notin \text{sub}(t)$.

(R.3) Again, $\mathcal{B}_a \subseteq \mathsf{ATm} \times \mathsf{Fm}$ and therefore $(t \cdot n_P, F) \notin \mathcal{B}_a$.

□

7. System $\mathsf{JN_V}$

Definition 7.26 (CS-Models). For a constant specification CS, a model $\mathcal{M} = (\mathsf{v}, \mathcal{B})$ is called a CS-*model* if $\mathsf{CS} \subseteq \mathcal{B}$.

A formula F is called *valid w.r.t. intermediate* CS-*models* (notation: $\Vdash_i^{\mathsf{CS}} F$) iff $\mathcal{M} \Vdash F$ for all intermediate CS-models $\mathcal{M}$.

A formula F is called *valid w.r.t. atomic* CS-*models* (notation: $\Vdash_a^{\mathsf{CS}} F$) iff $\mathcal{M} \Vdash F$ for all atomic CS-models $\mathcal{M}$.

8. Soundness and completeness for $\mathsf{JN_V}$

8.1. Soundness for $\mathsf{JN_V}$

Before proving soundness, we will need a couple of lemmas.

Lemma 8.1. *For an intermediate model* $\mathcal{M} = (\mathsf{v}, \mathcal{B})$ *and any atomic term* $t \in \mathsf{ATm}$,
$$(t, F) \in \mathcal{B} \quad \Leftrightarrow \quad (t, F) \in \mathcal{E}(\mathcal{B})$$

Proof. Direction from left to right is true for any model and follows from Lemma 7.16.

Direction from right to left is proved by induction on buildup of $\mathcal{E}(\mathcal{B})$:

- Basis: $(t, F) \in \mathcal{B}$, then the claim is trivial.

- Application: $(t \cdot s, F) \in \mathcal{E}(\mathcal{B})$.
  This case does not apply, since $t \cdot s \notin \mathsf{ATm}$.

- Substitution: $(\sigma t, F) \in \mathcal{E}(\mathcal{B})$ for $\sigma = \langle i, P, s \rangle$, for $(t, F) \in \mathcal{E}(\mathcal{B})$, $(s, P) \in \mathcal{E}(\mathcal{B})$.

  By assumption, $\sigma t \in \mathsf{ATm}$.

  By Lemma 7.7, $0 = \mathrm{cmp}(\sigma t) \geqslant \mathrm{cmp}(t) \geqslant 0$, therefore $\mathrm{cmp}(t) = 0$ and $t \in \mathsf{ATm}$.

  Since t is atomic, by definition of the basic substitution σ we have
  $$\sigma t = \begin{cases} s, & \text{if } t = n_P \text{ and } i = 0 \\ t, & \text{otherwise} \end{cases}$$

    – In the first case, $t = n_P \in \mathsf{ATm}$ and we know that $(t, F) \in \mathcal{B}$
      So we have $(t, F) = (n_P, F) \in \mathcal{B}$ for an intermediate model $\mathcal{M} = (\mathsf{v}, \mathcal{B})$.

73

8. Soundness and completeness for JN$_V$

By (R.1), $F = P$. Therefore, $(\sigma t, F) = (s, P)$, and in particular $s = \sigma t \in \mathsf{ATm}$, so induction hypothesis applies to (s, P) and we get $(\sigma t, F) = (s, P) \in \mathcal{B}$.

– In the second case, $t = \sigma t \in \mathsf{ATm}$, so induction hypothesis applies to (t, F) and we get $(\sigma t, F) = (t, F) \in \mathcal{B}$.

$\square$

Lemma 8.2. *For any normal terms t, s and any $i \in \mathbb{N}, P \in \mathsf{V}$, $\langle i, P, s \rangle t$ is also normal.*

Proof. Fix s. Proof by induction on t.

Base: t is atomic. Therefore, for $\sigma = \langle i, P, s \rangle$,

$$\sigma t = \begin{cases} s & t = n_P \text{ and } i = 0 \\ t & \text{otherwise.} \end{cases}$$

In either case the claim follows by normality of t or s.

Step: $t = u \cdot v$.
By choice of t, u and v are also normal, since $\mathrm{sub}(u) \cup \mathrm{sub}(v) \subseteq \mathrm{sub}(t)$. This means that the induction hypothesis applies for u and v. In particular, for $\sigma = \langle i, P, s \rangle$ follows that σu and $(\sigma {\downarrow} u)v$ are normal.

We have

$$\sigma t = \begin{cases} (\sigma u) \cdot v & i < \mathrm{num}_P(u) \\ u \cdot ((\sigma {\downarrow} u)v) & \text{otherwise.} \end{cases}$$

(a) $\sigma t = \sigma u \cdot v$.

Suppose σt is not normal, i.e. for some Q, r:

$$(n_Q \cdot r) \in \mathrm{sub}(\sigma t) = \{\sigma u \cdot v\} \cup \mathrm{sub}(\sigma u) \cup \mathrm{sub}(v)$$

As shown before, σu and v are normal. Therefore, $n_Q \cdot r = \sigma u \cdot v$ and $n_Q = \sigma u, r = v$.

By Lemma 7.7,

$$0 \leqslant \mathrm{cmp}(u) \leqslant \mathrm{cmp}(\sigma u) = \mathrm{cmp}(n_Q) = 0$$

Therefore, $\text{cmp}(u) = 0$ meaning that u must be atomic.

Since t is normal, $t = u \cdot v \neq n_P \cdot r$ for any r, which means that $u \neq n_P$.

Therefore, $\sigma u = u$ and $\sigma t = t$. Contradiction with the assumption that σt is not normal.

(b) $\sigma t = u \cdot ((\sigma \downarrow u)v)$.

Suppose σt is not normal, i.e. for some Q, r:
$$(n_Q \cdot r) \in \text{sub}(\sigma t) = \{u \cdot ((\sigma \downarrow u)v)\} \cup \text{sub}(u) \cup \text{sub}((\sigma \downarrow u)v)$$

As shown before, u and $(\sigma \downarrow u)v$ are normal.

Therefore, $n_Q \cdot r = u \cdot ((\sigma \downarrow u)v)$ and $n_Q = u, r = (\sigma \downarrow u)v$.

But that implies that $t = n_Q \cdot v$, which contradicts normality of t.

$\square$

Theorem 8.3. *Logic* $\mathsf{JN}^\mathsf{V}_\mathsf{CS}$ *is sound w.r.t. intermediate CS-models, i.e.*
$$\vdash_{\mathsf{JN}^\mathsf{V}_\mathsf{CS}} F \quad \Rightarrow \quad \Vdash^\mathsf{CS}_i F$$

Proof. Proof by induction on the length of derivation.
Let $\mathcal{M}_i = (\mathsf{v}, \mathcal{B})$ be an arbitrary intermediate CS-model.

(a) (Taut). All instances of propositional tautologies hold under all models, as the propositional part of truth relation is the same as truth-table semantics for classical propositional logic.

(b) (App). Assume $\mathcal{M}_i \Vdash t : (F \to G) \wedge s : F$.

From the definition of $\Vdash$ we have $\{(t, F \to G), (s, F)\} \subseteq \mathcal{E}(\mathcal{B})$.

From Lemma 7.16 follows that $(t \cdot s, G) \in \mathcal{E}(\mathcal{B})$, or equivalently that $\mathcal{M}_i \Vdash t \cdot s : G$.

Therefore, the implication holds for the model:
$$\mathcal{M}_i \Vdash t : (F \to G) \wedge s : F \to t \cdot s : G$$

8. Soundness and completeness for JN$_V$

(c) (N$_+$). By (R.1), $(n_P, F) \in \mathcal{B} \Leftrightarrow F = P$.

Therefore, $(n_P, P) \in \mathcal{B}$.

From Lemma 7.16, $(n_P, P) \in \mathcal{E}(\mathcal{B})$, and therefore $\mathcal{M}_i \Vdash n_P : P$.

(d) (N$_-$). Suppose $F \neq P$.

By (R.1), $(n_P, F) \notin \mathcal{B}$.

Since n_P is atomic, by Lemma 8.1 $(n_P, F) \notin \mathcal{E}(\mathcal{B})$.

Therefore, $\mathcal{M}_i \nVdash n_P : F$.

(e) (N$_\triangleleft$). The claim is that $(t, F) \notin \mathcal{E}(\mathcal{B})$ for t such that $n_P \cdot s \in \text{sub}(t)$.

We will show that for any $(t, F) \in \mathcal{E}(\mathcal{B})$, $n_P \cdot s \notin \text{sub}(t)$ by induction on buildup of $\mathcal{E}(\mathcal{B})$.

We will show that for every $(t, F) \in \mathcal{E}(\mathcal{B})$, t is normal.

<u>Basis:</u> $(t, F) \in \mathcal{B}$, then the case is covered by (R.2).

<u>Application:</u> $t = r \cdot q$ while $\{(r, G \to F), (q, G)\} \subseteq \mathcal{E}(\mathcal{B})$.

By induction hypothesis we have r, q normal.

We need to show that $(n_P \cdot s) \notin \text{sub}(t) = \{r \cdot q\} \cup \text{sub}(r) \cup \text{sub}(q)$, therefore it is sufficient to show that $n_P \cdot s \neq r \cdot q$.

If the opposite was true, i.e. $n_P \cdot s = r \cdot q$, then we conclude $n_P = r$ and $(n_P, G \to F) \in \mathcal{E}(\mathcal{B})$.

But, since $G \to F \neq P$, $(n_P, G \to F) \notin \mathcal{E}(\mathcal{B})$ as shown above. Contradiction.

<u>Substitution:</u> $t = \sigma r$, where $\sigma = \langle i, Q, q \rangle$ and $\{(r, F), (q, Q)\} \subseteq \mathcal{E}(\mathcal{B})$

By induction hypothesis, r and q are normal. Lemma 8.2 applies, showing that t is normal.

(f) (N$_\triangleright$). We need to show $\mathcal{M}_i \Vdash t \cdot n_P : F \to t : (P \to F)$.

Assume $\mathcal{M}_i \Vdash t \cdot n_P : F$, or equivalently $(t \cdot n_P, F) \in \mathcal{E}(\mathcal{B})$. We need to show that $(t, P \to F) \in \mathcal{E}(\mathcal{B})$.

We show that if $(r, F) \in \mathcal{E}(\mathcal{B})$ is of the form $(t \cdot n_P, F)$, then it follows that $(t, P \to F) \in \mathcal{E}(\mathcal{B})$. Proof by induction on buildup of $\mathcal{E}(\mathcal{B})$.

<u>Base:</u> By (R.3), $(t \cdot n_P, F) \notin \mathcal{B}$, so the antecedent of the implication is always false for this case.

8.1. Soundness for JN$_V$

Application: $(t \cdot n_P, F) \in \mathcal{E}(\mathcal{B})$, and for some H we have $(t, H \to F)$ and (n_P, H) are in $\mathcal{E}(\mathcal{B})$. By Lemma 8.1, $(n_P, H) \in \mathcal{B}$, and by (R.1) $H = P$.

Therefore, $(t, P \to F) \in \mathcal{E}(\mathcal{B})$.

Substitution: $(t \cdot n_P, F) \in \mathcal{E}(\mathcal{B})$, $t \cdot n_P = \sigma q$ for $\sigma = \langle i, Q, s \rangle$ and $(q, F), (s, Q) \in \mathcal{E}(\mathcal{B})$, for which the induction hypothesis applies.

Cases depending on σq:

a) $q = n_Q$, $i = 0$. Then $t \cdot n_P = \sigma q = s$.

   Since $(n_Q, F) \in \mathcal{E}(\mathcal{B})$, by Lemma 8.1 and (R.1) $F = Q$.

   Induction hypothesis applies for $(s, Q) = (t \cdot n_P, F)$, which gives the required $(t, P \to F)$.

b) $q \in \mathsf{ATm}$ but $i > 0$ or $q \neq n_Q$. Then $t \cdot n_P = \sigma q = q$. However, this contradicts $q \in \mathsf{ATm}$, so this case does not apply.

c) $q = q_1 \cdot q_2$ and $\sigma q = \sigma q_1 \cdot q_2$.

   Since $\sigma q = t \cdot n_P$, $q_2 = n_P$ and $\sigma q_1 = t$.

   We have $(q_1 \cdot n_P, F)$ and the induction hypothesis applies, giving $(q_1, P \to F) \in \mathcal{E}(\mathcal{B})$. Together with $(s, Q) \in \mathcal{E}(\mathcal{B})$, by Lemma 7.16 $(\sigma q_1, P \to F) = (t, P \to F) \in \mathcal{E}(\mathcal{B})$.

d) $q = q_1 \cdot q_2$ and $\sigma q = q_1 \cdot (\sigma \downarrow q_1) q_2$.

   Since $\sigma q = t \cdot n_P$, $(\sigma \downarrow q_1) q_2 = n_P$.

   By Lemma 7.7, $\mathrm{cmp}(q_2) \leqslant \mathrm{cmp}(n_P) = 0$, so $q_2 \in \mathsf{ATm}$.

   Since for atomic q_2, $n_P = \langle j, Q, s \rangle q_2$ is either q_2 or s, we have two possibilities:

   - $q_2 = n_P$.

     Induction hypothesis applies for $(q, F) = (q_1 \cdot n_P, F)$, so $(q_1, P \to F) = (t, P \to F) \in \mathcal{E}(\mathcal{B})$

   - $s = n_P$.

     In this case $\sigma = \langle i, Q, n_P \rangle$ for some i.

     We have $(s, Q) = (n_P, Q) \in \mathcal{E}(\mathcal{B})$. As shown above, this means that $P = Q$.

     Therefore, $(q, F) = (q_1 \cdot q_2, F) = (t \cdot n_Q, F) = (t \cdot n_P, F)$.

     Induction hypothesis applies, giving $(t, P \to F) \in \mathcal{E}(\mathcal{B})$.

8. Soundness and completeness for JN$_V$

(g) (N$_\sigma$). We need to show $\mathcal{M}_i \Vdash t : F \wedge s : P \to \sigma t : F$ for $\sigma = \langle i, P, s \rangle$.
Equivalently, we need to show that $\{(t,F),(s,P)\} \subseteq \mathcal{E}(\mathcal{B})$ entails $(\langle i,P,s\rangle t, F) \in \mathcal{E}(\mathcal{B})$. This is proven in Lemma 7.16.

(h) (MP). Modus ponens preserves truth in a model.

(i) (AN). Since $\mathcal{M}$ is a CS-model, then by definition $\mathsf{CS} \subseteq \mathcal{B}$.
By Lemma 7.16, we obtain $\mathsf{CS} \subseteq \mathcal{E}(\mathcal{B})$, giving $\mathcal{M}_i \Vdash (c,F)$ for all $(c,F) \in \mathsf{CS}$.

$\square$

8.2. Completeness for JN$_V$

The goal of this section is to show that JN$_{\mathsf{CS}}^{\mathsf{V}}$ is complete w.r.t. intermediate models.

Definition 8.4 (Consistency). A set Φ of formulas is called *consistent* if JN$_{\mathsf{CS}}^{\mathsf{V}} \not\vdash \neg(A_1 \wedge \cdots \wedge A_n)$ for any finite subset $\{A_1, \ldots, A_n\} \subseteq \Phi$.

A set Φ is called *maximal consistent* if it is consistent, and no proper superset of Φ is.

Remark 8.5 (Lindenbaum Lemma). For every consistent set Φ, there is a maximal consistent set Φ' such that $\Phi \subseteq \Phi'$.

Lemma 8.6. *Let Φ be a maximal consistent set.*

- *Φ contains all instances of all axioms.*
- *Φ is closed under* (MP), (AN).
- *$\neg F \in \Phi \Leftrightarrow F \notin \Phi$*
- *$(F \to G) \in \Phi \Leftrightarrow (F \notin \Phi \text{ or } G \in \Phi)$*
- *$(F \leftrightarrow G) \in \Phi \Leftrightarrow (F \in \Phi \Leftrightarrow G \in \Phi)$*

Remark 4.3 and Lemma 8.6 are standard for (maximal) consistent sets; for proofs, see e.g. [KS16].

Definition 8.7 (Induced model). Let Φ be a maximal consistent set of formulas. We define its *induced model* as $\mathcal{M}_\Phi := (\mathsf{v}_\Phi, \mathcal{B}_\Phi)$, where

8.2. Completeness for JN$_V$

- $\mathsf{v}_\Phi := \Phi \cap \mathsf{Prop}$

- $\mathcal{B}_\Phi := \{(t, F) \mid t : F \in \Phi, t \neq s \cdot n_P \text{ for any } s, P\}$

Lemma 8.8. *For any maximal consistent set Φ, $\mathcal{M}_\Phi$ is an intermediate CS-model.*

Proof. We need to show that (R.1)—(R.3) hold.

(R.1) $(n_P, F) \in \mathcal{B}_\Phi \Leftrightarrow F = P$.

Since Φ is maximal consistent, by Lemma 8.6 all instances of axioms (N$_+$), (N$_-$) belong to it.

This gives $n_P : F \in \Phi \Leftrightarrow F = P$ and thus (R.1) by definition of $\mathcal{B}_\Phi$.

(R.2) $n_P \cdot s \notin \mathrm{sub}(t)$ for all $(t, F) \in \mathcal{B}_\Phi$

By Lemma 8.6, all instances of axioms (N$_\triangleleft$) belong to Φ.

As above, this means that $n_P \cdot s \in \mathrm{sub}(t) \Rightarrow (t, F) \notin \mathcal{B}_\Phi$.

(R.3) $(t \cdot n_P, F) \notin \mathcal{B}_\Phi$

Immediately from definition of $\mathcal{B}_\Phi$.

To show that $\mathcal{M}_\Phi$ is a CS-model, observe that Φ is closed under (AN) by Lemma 8.6. It means that for every $(c, F) \in \mathsf{CS}$, $c : F \in \Phi$. Since $c \neq t \cdot n_P$ for any t, P, we have $(c, F) \in \mathcal{B}_\Phi$ by definition of $\mathcal{B}_\Phi$. □

Lemma 8.9 (Canonical evidence).

$$(t, F) \in \mathcal{E}(\mathcal{B}_\Phi) \quad \Leftrightarrow \quad t : F \in \Phi$$

Proof.

($\Rightarrow$) We have $(t, F) \in \mathcal{E}(\mathcal{B}_\Phi)$. Induction on the buildup of $\mathcal{E}(\mathcal{B}_\Phi)$:

Base: $(t, F) \in \mathcal{B}_\Phi$, therefore by definition $t : F \in \Phi$.

App: $(t \cdot s, F) \in \mathcal{E}(\mathcal{B}_\Phi)$, and $\{(t, G \to F), (s, G)\} \subseteq \mathcal{E}(\mathcal{B}_\Phi)$.

Induction hypothesis applies for t and s: $\{(t : G \to F), (s : G)\} \subseteq \Phi$

Since Φ is maximal consistent, by Lemma 8.6,

$t : (G \to F) \to (s : G \to (t : (G \to F) \wedge s : G)) \in \Phi$ ((Prop) instance)

8. Soundness and completeness for JN$_V$

$$(t : (G \to F) \land s : G) \to t \cdot s : F \in \Phi \quad ((\text{App}) \text{ instance})$$

By Lemma 8.6 and applying (MP) several times, $t \cdot s : F \in \Phi$.
<u>Substitution:</u> $(\langle i, P, s\rangle t, F) \in \mathcal{E}(\mathcal{B}_\Phi)$, and $\{(t, F), (s, P)\} \subseteq \mathcal{E}(\mathcal{B}_\Phi)$. Induction hypothesis applies for t and s: $\{(t : F), (s : P)\} \subseteq \Phi$. Since Φ is maximal consistent, by Lemma 8.6,

$$t : F \to (s : P \to (t : F \land s : P)) \in \Phi \quad ((\text{Prop}) \text{ instance})$$

$$(t : F \land s : P) \to \langle i, P, s\rangle t : F \in \Phi \quad ((\text{N}_\sigma) \text{ instance})$$

By Lemma 8.6 and applying (MP) several times, $\langle i, P, s\rangle t : F \in \Phi$.

($\Leftarrow$) We have $t : F \in \Phi$. Proof proceeds by induction on t.

There are two cases:

(a) $t \neq s \cdot n_P$ for any s, P. This includes the base case.
In this case, by definition of $\mathcal{B}_\Phi$ we have $(t, F) \in \mathcal{B}_\Phi$.
By Lemma 7.16, $(t, F) \in \mathcal{E}(\mathcal{B}_\Phi)$.

(b) $t = s \cdot n_P$ for some s, P.
In this case, by Lemma 8.6,

$$s \cdot n_P : F \to s : (P \to F) \in \Phi \quad ((\text{N}_\triangleright) \text{ instance})$$

Applying (MP), we get $s : (P \to F) \in \Phi$, and by induction hypothesis $(s, (P \to F)) \in \mathcal{E}(\mathcal{B}_\Phi)$.
Since $\mathcal{M}_\Phi$ is an intermediate model, $(n_P, P) \in \mathcal{B}_\Phi$ by (R.1).
Finally, applying Lemma 7.16 we get $(s \cdot n_P, F) = (t, F) \in \mathcal{E}(\mathcal{B})$.

$\square$

Definition 8.10. We define the *rank* for a formula F, denoted as $\text{rk}(F)$, inductively as follows:

- $\text{rk}(P) := 0$ for $P \in \text{Prop}$
- $\text{rk}(\neg A) := \text{rk}(A) + 1$

8.2. Completeness for JN$_V$

- $\mathsf{rk}(A \to B) := \mathsf{rk}(A) + \mathsf{rk}(B) + 1$
- $\mathsf{rk}(t : A) := 0$

Note that this is distinct from the definition of rk in Definition 4.8.

Lemma 8.11 (Truth lemma).
$$\mathcal{M}_\Phi \Vdash F \quad \Leftrightarrow \quad F \in \Phi$$

Proof. Induction on $\mathsf{rk}(F)$:

(a) $F \in \mathsf{Prop}$. We have, by definition, $F \in \Phi \Leftrightarrow F \in \mathsf{v}_\Phi \Leftrightarrow \mathcal{M}_\Phi \Vdash F$.

(b) $F = \neg G$. Since Φ is maximal consistent, $\neg G \in \Phi \Leftrightarrow G \notin \Phi$ by Lemma 8.6.
$\mathsf{rk}(\neg G) > \mathsf{rk}(G)$, so induction hypothesis applies for G, and therefore $\neg G \in \Phi \Leftrightarrow \mathcal{M}_\Phi \nVdash G$.

(c) $F = G \to H$.
Since Φ is maximal consistent, $G \to H \in \Phi \Leftrightarrow (G \notin \Phi$ or $H \in \Phi)$ by Lemma 8.6.
$\mathsf{rk}(G \to H) > \mathsf{rk}(G)$, $\mathsf{rk}(G \to H) > \mathsf{rk}(H)$, so induction hypothesis applies for G and H, yielding
$$(G \notin \Phi \text{ or } H \in \Phi) \Leftrightarrow (\mathcal{M}_\Phi \nVdash G \text{ or } \mathcal{M}_\Phi \Vdash H)$$
By definition of $\Vdash$ and the above, $G \to H \in \Phi \Leftrightarrow \mathcal{M}_\Phi \Vdash G \to H$.

(d) $F = t : G$. By Lemma 8.9, $t : G \in \Phi \Leftrightarrow (t, G) \in \mathcal{E}(\mathcal{B}_\Phi)$.
By definition of $\Vdash$, that means $t : G \in \Phi \Leftrightarrow \mathcal{M}_\Phi \Vdash t : G$.

$\square$

Theorem 8.12 (Completeness). *Logic* $\mathsf{JN}^V_{\mathsf{CS}}$ *is complete w.r.t. intermediate* CS*-models, i.e.*
$$\Vdash^{\mathsf{CS}}_i F \quad \Rightarrow \quad \vdash_{\mathsf{JN}^V_{\mathsf{CS}}} F$$

8. Soundness and completeness for JN$_V$

Proof. Proof by contraposition. Assume that $\nvdash_{\mathsf{JN}^V_{CS}} F$. In that case, $\{\neg F\}$ is a consistent set.

It is contained in a maximal consistent set Φ, and by Lemma 8.11 we have $\mathcal{M}_\Phi \Vdash \neg F$, or equivalently $\mathcal{M}_\Phi \nVdash F$.

Therefore, there is a counter-model for F and $\nvDash^{CS}_i F$. $\square$

9. Finite model property for $\mathsf{JN_V}$

For the following proofs, we will require a variant of *finite model property* w.r.t. intermediate models.

Definition 9.1. We call a model $\mathcal{M} = (\mathsf{v}, \mathcal{B})$ *finite* if both v and $\mathcal{B}$ are finite.

We call a CS-model $\mathcal{M} = (\mathsf{v}, \mathcal{B} \cup \mathsf{CS})$ *almost finite* if both v and $\mathcal{B}$ are finite.

We want to prove a form of finite model property which states that any satisfiable formula can be satisfied in a finite model.

However, since CS might be infinite, and any CS-model must include the entire CS in the basis, we cannot achieve "true" finite model property. Instead, we'll have almost finite intermediate CS-models. Since CS can be considered "given", such almost finite models still contain only finite information in addition to CS.

Before we prove this, we need a couple more definitions.

Definition 9.2 (Subformulas)**.** For a formula F, we define the *set of subformulas* $\mathrm{subf}(F)$ inductively as follows:

- $\mathrm{subf}(P) = \{P\}$,
- $\mathrm{subf}(\neg F) = \{\neg F\} \cup \mathrm{subf}(F)$,
- $\mathrm{subf}(F \to G) = \{F \to G\} \cup \mathrm{subf}(F) \cup \mathrm{subf}(G)$,
- $\mathrm{subf}(t : F) = \{t : F\}$. Notice that this definition is "shallow" as this case does not include $\mathrm{subf}(F)$ or propositional variables that may occur in nominals in t.

For any set of evidence pairs $\mathcal{F} \subseteq \mathsf{Tm} \times \mathsf{Fm}$, define its set of subformulas as $\mathrm{subf}(\mathcal{F}) := \bigcup_{(t,F) \in \mathcal{F}} \mathrm{subf}(F)$.

Definition 9.3. For a formula F, we define

9. Finite model property for JN$_V$

- *relevant evidence* $\mathcal{E}_F := \{(t, G) \mid t : G \in \text{subf}(F)\}$
- *relevant propositional valuation* $v_F := \{P \mid P \in \text{subf}(F)\}$

Note that both defined sets are finite for any formula F.

Definition 9.4 (Model agreement).
For a formula F, two models $\mathcal{M} = (v, \mathcal{B})$ and $\mathcal{M}' = (v', \mathcal{B}')$ are said to

- *agree on relevant propositional valuation for F* iff $v \cap v_F = v' \cap v_F$,
- *agree on relevant evidence for F* iff $\mathcal{E}(\mathcal{B}) \cap \mathcal{E}_F = \mathcal{E}(\mathcal{B}') \cap \mathcal{E}_F$.

Lemma 9.5. *If two models $\mathcal{M} = (v, \mathcal{B})$ and $\mathcal{M}' = (v', \mathcal{B}')$ agree on relevant propositional valuation and relevant evidence for F, then they agree on the valuation of F: $\mathcal{M} \Vdash F \Leftrightarrow \mathcal{M}' \Vdash F$.*

Proof. By induction on $\mathcal{M} \Vdash F$, since it only depends on relevant propositional valuation and evidence. $\square$

Lemma 9.6. *For any bases $\mathcal{B}, \mathcal{B}' \subseteq \mathsf{Tm} \times \mathsf{Fm}$ we have:*

(a) $\mathcal{B} \subseteq \mathcal{B}' \Rightarrow \mathcal{E}(\mathcal{B}) \subseteq \mathcal{E}(\mathcal{B}')$

(b) $\mathcal{E}(\mathcal{E}(\mathcal{B})) = \mathcal{E}(\mathcal{B})$

Proof. (a) Assume $\mathcal{B} \subseteq \mathcal{B}'$.

We show that $\mathcal{E}(\mathcal{B}) \subseteq \mathcal{E}(\mathcal{B}')$ by induction on the buildup of $\mathcal{E}(\mathcal{B})$.

Base: $(t, F) \in \mathcal{B}$.

Therefore, $(t, F) \in \mathcal{B}'$, and by Lemma 7.16 $(t, F) \in \mathcal{E}(\mathcal{B}')$.

App: $(t \cdot s, F) \in \mathcal{E}(\mathcal{B})$, and induction hypothesis applies for $(t, G \to F)$ and (s, G), giving $\{(t, G \to F), (s, G)\} \subseteq \mathcal{E}(\mathcal{B}')$.

From that and Lemma 7.16 we conclude $(t \cdot s, F) \in \mathcal{E}(\mathcal{B}')$.

Substitution: $(\sigma t, F) \in \mathcal{E}(\mathcal{B})$ for $\sigma = \langle i, P, s \rangle$, and induction hypothesis applies for (t, F) and (s, P) giving $\{(t, F), (s, P)\} \subseteq \mathcal{E}(\mathcal{B}')$.

From that and Lemma 7.16 we conclude $(\sigma t, F) \in \mathcal{E}(\mathcal{B}')$.

(b) By Lemma 7.16 we have $\mathcal{E}(\mathcal{B}) \subseteq \mathcal{E}(\mathcal{E}(\mathcal{B}))$.

Therefore, we only need to show $\mathcal{E}(\mathcal{E}(\mathcal{B})) \subseteq \mathcal{E}(\mathcal{B})$. We show that by induction on the buildup of $\mathcal{E}(\mathcal{E}(\mathcal{B}))$.

Base: $(t, F) \in \mathcal{E}(\mathcal{B})$, and that shows the claim immediately.

App: $(t \cdot s, F) \in \mathcal{E}(\mathcal{E}(\mathcal{B}))$, and induction hypothesis applies for (s, G) and $(t, G \to F)$, giving $\{(t, G \to F), (s, G)\} \subseteq \mathcal{E}(\mathcal{B})$.

From that and Lemma 7.16 we conclude $(t \cdot s, F) \in \mathcal{E}(\mathcal{B})$.

Substitution: $(\sigma t, F) \in \mathcal{E}(\mathcal{E}(\mathcal{B}))$ for $\sigma = \langle i, P, s \rangle$, and induction hypothesis applies for (t, F) and (s, P) giving $\{(t, F), (s, P)\} \subseteq \mathcal{E}(\mathcal{B})$.

From that and Lemma 7.16 we conclude $(\sigma t, F) \in \mathcal{E}(\mathcal{B})$. □

Lemma 9.7. *For any basis $\mathcal{B}$ and any (t, F) such that $(t, F) \in \mathcal{E}(\mathcal{B})$, there is a finite subset $\mathcal{B}' \subseteq \mathcal{B}$ such that $(t, F) \in \mathcal{E}(\mathcal{B}')$.*

Proof. Proof by induction on the buildup of $\mathcal{E}(\mathcal{B})$.

Base: $(t, F) \in \mathcal{B}$.
Then we can take $\mathcal{B}' := \{(t, F)\}$. By Lemma 7.16, $(t, F) \in \mathcal{B}' \subseteq \mathcal{E}(\mathcal{B}')$.

App: $(t \cdot s, F) \in \mathcal{E}(\mathcal{B})$, and $\{(t, G \to F), (s, G)\} \subseteq \mathcal{E}(\mathcal{B})$.

Induction hypothesis applies, so there exist finite subsets $\mathcal{B}'_t, \mathcal{B}'_s$ of $\mathcal{B}$ such that
$$(t, G \to F) \in \mathcal{E}(\mathcal{B}'_t) \quad (s, G) \in \mathcal{E}(\mathcal{B}'_s)$$

Take $\mathcal{B}' := \mathcal{B}'_t \cup \mathcal{B}'_s$. Since both sets in the union are finite, $\mathcal{B}'$ is finite.

By Lemma 9.6, $\{(t, G \to F), (s, G)\} \subseteq \mathcal{E}(\mathcal{B}')$. From Lemma 7.16 follows the required $(t \cdot s, F) \in \mathcal{E}(\mathcal{B}')$.

Substitution: Analogous to the above case. □

Note that the above proof gives a specific construction for a finite subset of $\mathcal{B}$ that is enough to provide evidence for an evidence pair (t, F).

Definition 9.8. For any basis $\mathcal{B}$ and any (t, F), define $\mathcal{B}_{(t,F)}$ as the subset of $\mathcal{B}$ provided by the Lemma 9.7 if $(t, F) \in \mathcal{E}(\mathcal{B})$, and $\varnothing$ otherwise.

Definition 9.9. For an intermediate CS-model $\mathcal{M} = (\mathsf{v}, \mathcal{B})$ and a formula F, define $\mathrm{Fin}(\mathcal{B}, F)$ as follows:

$$\mathrm{Fin}(\mathcal{B}, F) := \left(\bigcup_{(t, G) \in \mathcal{E}_F} \mathcal{B}_{(t, G)} \right) \cup \{(n_P, P) \mid P \in \mathsf{V}\}$$

9. Finite model property for JN$_V$

Theorem 9.10 (Finite model property I). *For every intermediate model $\mathcal{M}$ and formula F, the model $\mathcal{M}' := (\mathsf{v} \cap \mathsf{v}_F, \mathrm{Fin}(\mathcal{B}, F))$ has the following properties:*

(a) *$\mathcal{M}'$ is a finite model.*

(b) *$\mathcal{M}'$ agrees with $\mathcal{M}$ on relevant propositional valuation for F.*

(c) *$\mathcal{M}'$ agrees with $\mathcal{M}$ on relevant evidence for F.*

(d) *$\mathcal{M}' \Vdash F \Leftrightarrow \mathcal{M} \Vdash F$.*

(e) *$\mathcal{M}'$ is an intermediate model.*

Proof. We show claims one by one.

(a) *$\mathcal{M}'$ is a finite model.*

v_F is finite for every formula F, therefore $\mathsf{v}' := \mathsf{v} \cap \mathsf{v}_F$ is finite.

$\{(n_P, P) \mid P \in \mathsf{V}\}$ is finite because V is finite, $\mathcal{E}_F$ is finite for any formula F, and $\mathcal{B}_{(t,G)}$ is finite by construction.

Therefore, $\mathcal{B}' := \mathrm{Fin}(\mathcal{B}, F)$ is finite as a union of finitely many finite sets.

(b) *$\mathcal{M}'$ agrees with $\mathcal{M}$ on relevant propositional valuation for F.*

$\mathsf{v}' = \mathsf{v} \cap \mathsf{v}_F$, therefore $\mathsf{v}' \cap \mathsf{v}_F = (\mathsf{v} \cap \mathsf{v}_F) \cap \mathsf{v}_F = \mathsf{v} \cap \mathsf{v}_F$, so the models agree on propositional valuation.

(c) *$\mathcal{M}'$ agrees with $\mathcal{M}$ on relevant evidence for F.*

Observe the following:

- $\mathcal{B}' \subseteq \mathcal{B}$.

  All of $\mathcal{B}_{(t,G)}$ are subsets of $\mathcal{B}$ by construction.

  Since $\mathcal{M}$ is intermediate, it satisfies (R.1), which means that $\{(n_P, P) \mid P \in \mathsf{V}\} \subseteq \mathcal{B}$.

  Therefore, $\mathcal{B}'$ is as subset of $\mathcal{B}$ as a union of subsets of $\mathcal{B}$.

- $\mathcal{E}(\mathcal{B}') \cap \mathcal{E}_F \subseteq \mathcal{E}(\mathcal{B})$.

  From the above and Lemma 9.6, we have $\mathcal{E}(\mathcal{B}') \subseteq \mathcal{E}(\mathcal{B})$, which is stronger than the claim.

- $\mathcal{E}(\mathcal{B}) \cap \mathcal{E}_F \subseteq \mathcal{E}(\mathcal{B}')$.

  Suppose $(t, G) \in \mathcal{E}(\mathcal{B}) \cap \mathcal{E}_F$.

  Then $\mathcal{B}_{(t,G)} \subseteq \mathcal{B}'$ by construction of $\mathcal{B}'$, and $\mathcal{B}_{(t,G)}$ is non-empty. By construction of $\mathcal{B}_{(t,G)}$, $(t, G) \in \mathcal{E}(\mathcal{B}_{(t,G)})$.

  Finally, by Lemma 9.6, $\mathcal{E}(\mathcal{B}_{(t,G)}) \subseteq \mathcal{E}(\mathcal{B}')$, giving $(t, G) \in \mathcal{E}(\mathcal{B}')$.

- $\mathcal{E}(\mathcal{B}) \cap \mathcal{E}_F = \mathcal{E}(\mathcal{B}') \cap \mathcal{E}_F$.

  Follows directly from the last two claims.

The above shows that $\mathcal{M}$ and $\mathcal{M}'$ agree on relevant evidence.

(d) $\mathcal{M}' \Vdash F \Leftrightarrow \mathcal{M} \Vdash F$.

Since the two models agree on relevant propositional valuation and evidence, claim follows by Lemma 9.5.

(e) $\mathcal{M}'$ *is an intermediate model.*

We need to show (R.1) — (R.3) for $\mathcal{B}'$.

(R.1) $(n_P, F) \in \mathcal{B}' \Leftrightarrow F = P$ (for $P \in \mathsf{V}$)

Since $\{(n_P, P) \mid P \in \mathsf{V}\} \subseteq \mathcal{B}'$ by construction, the direction from right to left is valid.

$\mathcal{M}$ is an intermediate model, so (R.1) applies for $\mathcal{B}$.

Suppose $(n_P, F) \in \mathcal{B}'$ for $P \in \mathsf{V}$. Since $\mathcal{B}' \subseteq \mathcal{B}$, $(n_P, F) \in \mathcal{B}$ and therefore $F = P$.

(R.2) $n_P \cdot s \notin \mathrm{sub}(\mathcal{B}')$

$\mathcal{M}$ is intermediate; therefore, (R.2) applies for $\mathcal{B}$ and we conclude $n_P \cdot s \notin \mathrm{sub}(\mathcal{B})$.

$\mathcal{B}' \subseteq \mathcal{B}$ and consequently $\mathrm{sub}(\mathcal{B}') \subseteq \mathrm{sub}(\mathcal{B})$.

Therefore, we obtain $n_P \cdot s \notin \mathrm{sub}(\mathcal{B}')$.

(R.3) $(t \cdot n_P, F) \notin \mathcal{B}'$

Analogous to the above case.

□

9. Finite model property for JN$_V$

Theorem 9.10 gives us a model that may not be a CS-model. If we want to show a proper almost finite model property, we need to add constant specification:

Theorem 9.11 (Finite model property II). *For every intermediate CS-model $\mathcal{M}$ and formula F, the model $\mathcal{M}' := (\mathsf{v} \cap \mathsf{v}_F, \mathrm{Fin}(\mathcal{B}, F) \cup \mathsf{CS})$ has the following properties:*

(a) $\mathcal{M}'$ is a CS-model.

(b) $\mathcal{M}'$ is an almost finite model.

(c) $\mathcal{M}'$ agrees with $\mathcal{M}$ on relevant propositional valuation for F.

(d) $\mathcal{M}'$ agrees with $\mathcal{M}$ on relevant evidence for F.

(e) $\mathcal{M}' \Vdash F \Leftrightarrow \mathcal{M} \Vdash F$.

(f) $\mathcal{M}'$ is an intermediate model.

Proof. Most of the proof is based on facts established in Theorem 9.10.

(a) $\mathcal{M}'$ is a CS-*model*.

By definition, since $\mathsf{CS} \subseteq \mathrm{Fin}(\mathcal{B}, F) \cup \mathsf{CS}$.

(b) $\mathcal{M}'$ is an almost finite model.

In Theorem 9.10, we established that $\mathsf{v} \cap \mathsf{v}_F$ and $\mathrm{Fin}(\mathcal{B}, F)$ are finite; this exactly means that $\mathcal{M}'$ is almost finite.

(c) $\mathcal{M}'$ agrees with $\mathcal{M}$ on relevant propositional valuation for F.

Shown in Theorem 9.10.

(d) $\mathcal{M}'$ agrees with $\mathcal{M}$ on relevant evidence for F.

The proof follows the same reasoning as in Theorem 9.10, with a minor addition in the proof of sub-claim $\mathcal{B}' \subseteq \mathcal{B}$.

Theorem 9.10 shows that $\mathrm{Fin}(\mathcal{B}, F) \subseteq \mathcal{B}$, and since $\mathcal{M}$ is a CS-model, $\mathsf{CS} \subseteq \mathcal{B}$ as well.

Together, this gives $\mathrm{Fin}(\mathcal{B}, F) \cup \mathsf{CS} = \mathcal{B}' \subseteq \mathcal{B}$

(e) $\mathcal{M}' \Vdash F \Leftrightarrow \mathcal{M} \Vdash F$.

As before, follows from two previous claims and Lemma 9.5.

(f) $\mathcal{M}'$ *is an intermediate model.*

Same as in Theorem 9.10.

$\square$

10. Atomic model semantics for JN$_V$

The goal of this chapter is to show that atomic models give a sound and complete axiomatization for JN$_V$.

By Lemma 7.25, any atomic model is also an intermediate model; therefore, we automatically get soundness. Instead of directly proving completeness, we will instead prove that any formula that has an intermediate counter-model also has an atomic counter-model — by transforming the initial counter-model while preserving validity of a given formula.

Starting from an arbitrary intermediate CS-model $\mathcal{M}$ for a formula F, Theorem 9.11 gives an almost finite intermediate CS-model for it. However, this model may not yet be atomic.

We can define a measure of "non-atomicity" of a model, or *complexity*, in terms of the number of applications in its basis. So, given an intermediate CS-model $\mathcal{M}$ we construct a finite model $\mathcal{M}_0$ with possibly non-zero complexity cmp($\mathcal{M}_0$).

The idea is to iteratively modify $\mathcal{M}_0$ in a way that reduces complexity while still producing models that agree on relevant evidence for F, which will ensure that validity is preserved by Lemma 9.5. A complexity of zero would mean that the model is atomic. Since complexity is finite, the process terminates.

However, we will need to impose a restriction on the constant specification if we want to produce a CS-model with this process (as opposed to an arbitrary model).

10. Atomic model semantics for JN_V

10.1. Preliminary definitions and lemmas

We define *complexity* as a measure of "non-atomicity" of a model.

Definition 10.1 (Basis and model complexity).

- We define *basis complexity* for a basis $\mathcal{B}$ as

$$\operatorname{cmp}(\mathcal{B}) := \sum_{(t,F) \in \mathcal{B}} \operatorname{cmp}(t)$$

if this sum is finite, and $\operatorname{cmp}(\mathcal{B}) := \infty$ otherwise.

- We define *model complexity* for a model $\mathcal{M} = (v, \mathcal{B})$ as

$$\operatorname{cmp}(\mathcal{M}) := \operatorname{cmp}(\mathcal{B})$$

Remark 10.2. For any almost finite or finite model $\mathcal{M}$, its complexity $\operatorname{cmp}(\mathcal{M})$ is always finite.

During the atomization procedure, we will temporarily lose the full constant specification from the basis. In order to accurately track which parts of the constant specification are retained, we will need the notions of CS *fragments* and CS-*closure*.

Definition 10.3 (CS fragments).

- For a constant c, we define the *c-fragment of* CS as

$$\mathsf{CS}\!\restriction_c := \{(c', F) \mid (c', F) \in \mathsf{CS}, c' = c\}$$

- For a basis $\mathcal{B}$, we define the $\mathcal{B}$-*fragment of* CS as

$$\mathsf{CS}\!\restriction_{\mathcal{B}} := \{(c, F) \mid (c, F) \in \mathsf{CS}, c \in \operatorname{sub}(\mathcal{B})\} = \bigcup_{c \in \operatorname{sub}(\mathcal{B})} \mathsf{CS}\!\restriction_c$$

Definition 10.4 (CS-closed model, CS-closure).

- A model $\mathcal{M} = (v, \mathcal{B})$ is called CS-*closed* if its $\mathcal{B}$ basis contains the $\mathcal{B}$-fragment of CS:

$$\mathsf{CS}\!\restriction_{\mathcal{B}} \subseteq \mathcal{B}$$

10.1. Preliminary definitions and lemmas

- For a model $\mathcal{M} = (\mathsf{v}, \mathcal{B})$, define its CS-*closure* as

$$\mathcal{M}^{\mathsf{CS}} := (\mathsf{v}, \mathcal{B} \cup \mathsf{CS}\!\restriction_{\mathcal{B}})$$

Lemma 10.5. *A CS-closure of a model is a CS-closed model.*

Proof. Take a model $\mathcal{M} = (\mathsf{v}, \mathcal{B})$.
Then its CS-closure is $\mathcal{M}^{\mathsf{CS}} = (\mathsf{v}, \mathcal{B} \cup \mathsf{CS}\!\restriction_{\mathcal{B}})$.
We need to show that for every $c \in \mathrm{sub}(\mathcal{B} \cup \mathsf{CS}\!\restriction_{\mathcal{B}})$, $\mathsf{CS}\!\restriction_c \subseteq \mathcal{B} \cup \mathsf{CS}\!\restriction_{\mathcal{B}}$.
We have $\mathrm{sub}(\mathcal{B} \cup \mathsf{CS}\!\restriction_{\mathcal{B}}) = \mathrm{sub}(\mathcal{B}) \cup \mathrm{sub}(\mathsf{CS}\!\restriction_{\mathcal{B}})$. we have two cases:

(a) $c \in \mathrm{sub}(\mathcal{B})$.

Then, by definition of $\mathsf{CS}\!\restriction_{\mathcal{B}}$ we have $\mathsf{CS}\!\restriction_c \subseteq \mathsf{CS}\!\restriction_{\mathcal{B}} \subseteq \mathcal{B} \cup \mathsf{CS}\!\restriction_{\mathcal{B}}$.

(b) $c \in \mathrm{sub}(\mathsf{CS}\!\restriction_{\mathcal{B}})$.

We have

$$\mathrm{sub}(\mathsf{CS}\!\restriction_{\mathcal{B}}) = \bigcup_{(t,F) \in \mathsf{CS}\restriction_{\mathcal{B}}} \mathrm{sub}(t) = \bigcup_{c \in \mathrm{sub}(\mathcal{B})} \bigcup_{(c,F) \in \mathsf{CS}} \mathrm{sub}(c)$$
$$= \bigcup_{c \in \mathrm{sub}(\mathcal{B})} \bigcup_{(c,F) \in \mathsf{CS}} \{c\} \subseteq \mathrm{sub}(\mathcal{B}).$$

Therefore, this c is already covered by the previous case.

$\square$

Definition 10.6 (Relevant CS). For a formula F, we define *relevant constant specification* CS_F as the $\mathcal{E}_F$-fragment of CS:

$$\mathsf{CS}_F := \mathsf{CS}\!\restriction_{\mathcal{E}_F} = \{(c, G) \mid (c, G) \in \mathsf{CS}, c \in \mathrm{sub}(\mathcal{E}_F)\}$$

We say that a model $\mathcal{M} = (\mathsf{v}, \mathcal{B})$ *contains* CS_F if $\mathsf{CS}_F \subseteq \mathcal{B}$.

Lemma 10.7. *For any model $\mathcal{M} = (\mathsf{v}, \mathcal{B})$ and any formula F,*

$$F \notin \mathrm{subf}(\mathcal{B}) \quad \Rightarrow \quad F \notin \mathrm{subf}(\mathcal{E}(\mathcal{B}))$$

Proof. Assuming $F \notin \mathrm{subf}(\mathcal{B})$, we show by induction on buildup of $\mathcal{E}(\mathcal{B})$ that $F \notin \mathrm{subf}(G)$ for any $(t, G) \in \mathcal{E}(\mathcal{B})$.

10. Atomic model semantics for JN$_V$

Base: $(t, G) \in \mathcal{B}$.
Then subf$(G) \subseteq$ subf$(\mathcal{B})$ and therefore $F \notin$ subf(G) by assumption.
App: $(t \cdot s, G) \in \mathcal{E}(\mathcal{B})$, and $\{(t, H \to G), (s, H)\} \subseteq \mathcal{E}(\mathcal{B})$.
Induction hypothesis applies, so $F \notin$ subf$(H \to G)$.
Therefore, $F \notin$ subf$(G) \subseteq$ subf$(H \to G)$.
Substitution: $(\langle i, P, s \rangle t, G) \in \mathcal{E}(\mathcal{B})$, and $\{(t, G), (s, P)\} \subseteq \mathcal{E}(\mathcal{B})$.
Directly by induction hypothesis, $F \notin$ subf(G). □

In order to preserve validity of the formula on the last step of atomization, we will need a restriction on constant specification, *local finiteness*.

Definition 10.8 (Locally finite CS). *A constant specification is said to be locally finite if, for every constant c, the c-fragment of CS is finite.*

As a remark, locally finite CS can be axiomatically appropriate, but not schematic (for definitions, see e.g. [Fit05]).

We will need to explicitly work with compositions of basic substitutions. Here, we introduce *substitutions*, *substitution sets*, and several operations on substitutions:

Definition 10.9 (Substitutions). We define *substitutions* as the free monoid $\mathcal{S}^*$ on the set of all basic substitutions $\mathcal{S}$ with λ denoting the empty word. To clarify notation, we will use $\sigma_1 \circ \sigma_2$ to denote concatenation.
For a substitution $\Sigma \in \mathcal{S}^*$, $\Sigma[t]$ is defined inductively as follows:

- $\lambda[t] := t$ (so λ is the identity substitution)

- $(\Sigma' \circ \sigma)[t] := \Sigma'[\sigma t]$

Remark 10.10. For any two substitutions Σ_1, Σ_2 and any term t,

$$(\Sigma_1 \circ \Sigma_2)[t] = \Sigma_1[\Sigma_2[t]]$$

Definition 10.11 (Substitution sets). For a basis $\mathcal{B}$, define its *basic substitution set* $\mathcal{S}(\mathcal{B})$ as all basic substitutions that can appear in the buildup of $\mathcal{E}(\mathcal{B})$:

$$\mathcal{S}(\mathcal{B}) := \{\langle i, P, s \rangle \mid i \in \mathbb{N}, P \in V, (s, P) \in \mathcal{E}(\mathcal{B})\}$$

For a basis $\mathcal{B}$, define its *full substitution set* as the free monoid $\mathcal{S}(\mathcal{B})^*$ on its basic substitution set $\mathcal{S}(\mathcal{B})$.

10.1. Preliminary definitions and lemmas

Lemma 10.12. *For any basis $\mathcal{B}$, $(t, F) \in \mathcal{E}(\mathcal{B})$ and $\Sigma \in \mathcal{S}(\mathcal{B})^*$, we have $(\Sigma[t], F) \in \mathcal{E}(\mathcal{B})$.*

Proof. Induction on Σ for all $(t, F) \in \mathcal{E}(\mathcal{B})$.
Base: $\Sigma = \lambda$. Then $(\Sigma[t], F) = (t, F) \in \mathcal{E}(\mathcal{B})$.
Step: $\Sigma = \Sigma' \circ \sigma$ for $\sigma = \langle i, P, s \rangle$, and by induction hypothesis we have $(\Sigma'[t], F) \in \mathcal{E}(\mathcal{B})$ for all $(t, F) \in \mathcal{E}(\mathcal{B})$.
Take an arbitrary pair $(r, G) \in \mathcal{E}(\mathcal{B})$. We have $(\Sigma[r], G) = (\Sigma'[\sigma r], G)$.
Since $\Sigma \in \mathcal{S}(\mathcal{B})^*$, we infer that $\sigma \in \mathcal{S}(\mathcal{B})$ and therefore $(s, P) \in \mathcal{E}(\mathcal{B})$.
By Lemma 7.16 we obtain $(\sigma r, G) \in \mathcal{E}(\mathcal{B})$.
Therefore, by induction hypothesis $(\Sigma'[\sigma r], G) \in \mathcal{E}(\mathcal{B})$. □

Definition 10.13 (Substitution lifting). For a substitution Σ, we define *lifting of Σ by t* inductively as follows:

- $(\lambda \uparrow t) := \lambda$

- $((\Sigma' \circ \langle i, P, s \rangle) \uparrow t) := (\Sigma' \uparrow t) \circ \langle i + \operatorname{num}_P(t), P, s \rangle$

Lemma 10.14. *For any substitution Σ and terms t, s,*

$$(\Sigma \uparrow t)[t \cdot s] = t \cdot \Sigma[s]$$

Proof. Proof by induction on Σ for arbitrary s.
Base: $\Sigma = \lambda$.
By applying the definition of substitutions and filtering, we observe the following sequence of equivalent reductions of the claim:

$$(\lambda \uparrow t)[t \cdot s] = t \cdot \lambda[s]$$
$$\lambda[t \cdot s] = t \cdot s$$
$$t \cdot s = t \cdot s$$

Step: $\Sigma = \Sigma' \circ \sigma$ for $\sigma = \langle i, P, q \rangle$
By definition, $\Sigma \uparrow t = (\Sigma' \uparrow t) \circ \langle i + \operatorname{num}_P(t), P, q \rangle$.

Observe that $i + \mathrm{num}_P(t) \geqslant \mathrm{num}_P(t)$, therefore we get

$$\begin{aligned}
(\Sigma \uparrow t)[t \cdot s] &= ((\Sigma' \uparrow t) \circ \langle i + \mathrm{num}_P(t), P, q \rangle)[t \cdot s] \\
&= (\Sigma' \uparrow t)[\langle i + \mathrm{num}_P(t), P, q \rangle (t \cdot s)] \\
&= (\Sigma' \uparrow t)[t \cdot \langle i + \mathrm{num}_P(t) - \mathrm{num}_P(t), P, q \rangle s] \\
&= (\Sigma' \uparrow t)[t \cdot \langle i, P, q \rangle s] \\
&= (\Sigma' \uparrow t)[t \cdot \sigma s]
\end{aligned}$$

By induction hypothesis for Σ' and $s' := \sigma s$,

$$(\Sigma' \uparrow t)[t \cdot (\sigma s)] = t \cdot \Sigma'[\sigma s]$$

Using all of the above, we obtain the claim:

$$\begin{aligned}
(\Sigma \uparrow t)[t \cdot s] &= (\Sigma' \uparrow t)[t \cdot \sigma s] \\
&= t \cdot \Sigma'[\sigma s] \\
&= t \cdot (\Sigma' \circ \sigma)[s] \\
&= t \cdot \Sigma[s]
\end{aligned}$$

$\square$

Definition 10.15 (Substitution filtering). For a substitution Σ, we define *filtering of Σ by t* inductively as follows:

- $\lambda|_t := \lambda$

- $(\Sigma' \circ \langle i, P, s \rangle)|_t := \begin{cases} \Sigma'|_{\langle i, P, s \rangle t} \circ \langle i, P, s \rangle, & i < \mathrm{num}_P(t) \\ \Sigma'|_t, & i \geqslant \mathrm{num}_P(t) \end{cases}$

Intuitively, this discards basic substitutions that "do not apply" in the sense of Lemma 7.5.

Remark 10.16. For any basis $\mathcal{B}$ and any term t,

$$\Sigma \in \mathcal{S}(\mathcal{B})^* \quad \Rightarrow \quad \{(\Sigma \uparrow t), \Sigma|_t\} \subseteq \mathcal{S}(\mathcal{B})^*$$

Lemma 10.17. *For any substitution Σ and terms t, s,*

$$\Sigma|_t[t \cdot s] = \Sigma[t] \cdot s$$

10.1. Preliminary definitions and lemmas

Proof. Proof by induction on Σ for arbitrary t.

Base: $\Sigma = \lambda$.

By applying the definition of substitutions and filtering, we observe the following sequence of equivalent reductions of the claim:

$$\lambda|_t[t \cdot s] = \lambda[t] \cdot s$$
$$\lambda[t \cdot s] = t \cdot s$$
$$t \cdot s = t \cdot s$$

Step: $\Sigma = \Sigma' \circ \sigma$.

Induction hypothesis: for any term t', $\Sigma'|_{t'}[t' \cdot s] = \Sigma'[t'] \cdot s$.

We need to show:

$$(\Sigma' \circ \sigma)|_t[t \cdot s] = (\Sigma' \circ \sigma)[t] \cdot s$$

Note that $(\Sigma' \circ \sigma)[t] \cdot s = \Sigma'[\sigma t] \cdot s$.

There are two possible cases depending on $\sigma = \langle i, P, r \rangle$ and t:

(a) $i < \text{num}_P(t)$. In this case,

$$(\Sigma' \circ \sigma)|_t[t \cdot s] = (\Sigma'|_{\sigma t} \circ \sigma)[t \cdot s] = \Sigma'|_{\sigma t}[\sigma(t \cdot s)]$$

Since $i < \text{num}_P(t)$, we have $\sigma(t \cdot s) = \sigma t \cdot s$.

Using the above, the claim reduces to

$$\Sigma'|_{\sigma t}[\sigma t \cdot s] = \Sigma'[\sigma t] \cdot s,$$

which is exactly the induction hypothesis for $t' = \sigma t$.

(b) $i \geqslant \text{num}_P(t)$. In this case,

$$(\Sigma' \circ \sigma)|_t[t \cdot s] = \Sigma'|_t[t \cdot s]$$

Also, since $i \geqslant \text{num}_P(t)$, we have $\sigma t = t$ by Lemma 7.5.

Therefore, the claim reduces to

$$(\Sigma'|_t)[t \cdot s] = \Sigma'[t] \cdot s,$$

which is exactly the induction hypothesis for $t' = t$.

□

Lemma 10.18 (Combining substitutions). *For any substitutions Σ_1, Σ_2 and terms t, s, there exists a combining substitution $\mathrm{Comb}(\Sigma_1, t, \Sigma_2)$ such that*
$$\Sigma_1[t] \cdot \Sigma_2[s] = \mathrm{Comb}(\Sigma_1, t, \Sigma_2)[t \cdot s]$$
Moreover, for any basis $\mathcal{B}$,
$$\Sigma_1, \Sigma_2 \in \mathcal{S}(\mathcal{B})^* \quad \Rightarrow \quad \mathrm{Comb}(\Sigma_1, t, \Sigma_2) \in \mathcal{S}(\mathcal{B})^*$$

Proof. We claim that the required $\mathrm{Comb}(\Sigma_1, t, \Sigma_2)$ is $(\Sigma_2 \uparrow (\Sigma_1[t])) \circ \Sigma_1|_t$. Using Lemma 10.17, we observe that
$$((\Sigma_2 \uparrow (\Sigma_1[t])) \circ \Sigma_1|_t)[t \cdot s] = (\Sigma_2 \uparrow (\Sigma_1[t]))[\Sigma_1|_t[t \cdot s]]$$
$$= (\Sigma_2 \uparrow (\Sigma_1[t]))[\Sigma_1[t] \cdot s]$$

Applying Lemma 10.14 for Σ_2 and $(\Sigma_1[t])$ to the last expression, we obtain the claim:
$$(\Sigma_2 \uparrow (\Sigma_1[t]))[\Sigma_1[t] \cdot s] = \Sigma_1[t] \cdot \Sigma_2[s]$$

□

We will need the following observation regarding substitutions and constants:

Lemma 10.19. *For a basic substitution $\sigma = \langle i, P, s \rangle$ and a term t, one of the following is true:*

(a) $\sigma t = t$.

(b) *For every constant $c \in \mathrm{sub}(s) \cup \mathrm{sub}(t)$, $c \in \mathrm{sub}(\sigma t)$.*

Proof. Induction on t for arbitrary σ.

- Base: $t \in \mathsf{ATm}$. There are 2 sub-cases:
  - $t = n_P, i = 0$. Then $\sigma t = s$. In this case, $\mathrm{sub}(s) \cup \mathrm{sub}(t) = \mathrm{sub}(s) \cup \{n_P\}$. If a constant c is in $\mathrm{sub}(s) \cup \{n_P\}$, it must be in $\mathrm{sub}(s)$ as n_P is not a constant. But $c \in \mathrm{sub}(\sigma t) = \mathrm{sub}(s)$. This shows case (b) for this branch.

- $t \neq n_P$ or $i > 0$. Then $\sigma t = t$ which shows case (a).
- Step: $t = r \cdot q$. There are 2 sub-cases:
  - $i < \text{num}_P(r)$. Then $\sigma t = (\sigma r) \cdot q$.

    By induction hypothesis, either (a) or (b) is valid for σr.

    In case of (a) for σr, we have $\sigma t = \sigma r \cdot q = r \cdot q = t$, so we've shown case (a) for t.

    Suppose that case (b) applies for σr, and suppose that a constant c is in $\text{sub}(s) \cup \text{sub}(t)$.

    We have $\text{sub}(t) = \text{sub}(r) \cup \text{sub}(q) \cup \{r \cdot q\}$, and $r \cdot q$ is not a constant, therefore we have $c \in \text{sub}(r)$ or $c \in \text{sub}(q)$.

    We need to prove that
    $$c \in \text{sub}(\sigma t) = \text{sub}((\sigma r) \cdot q) = \text{sub}(\sigma r) \cup \text{sub}(q) \cup \{(\sigma r) \cdot q\}$$

    If $c \in \text{sub}(q)$ is the case, the above is true.

    If $c \in \text{sub}(r)$, remember that we assumed that case (b) applies for σr. From this we conclude that $c \in \text{sub}(\sigma r) \subseteq \text{sub}(\sigma t)$, and therefore (b) applies for σt.

  - $i \geq \text{num}_P(r)$. Then $\sigma t = r \cdot ((\sigma \downarrow r)q)$.

    Identical to the above, as induction hypothesis applies for an arbitrary σ.

□

Finally, we'll need a notion of a variable *fresh for a set of formulas*:

Definition 10.20 (Fresh variable). For a formula F, define $\text{PV}(F)$ as the set of propositional variables that occur in it. It is finite for any formula.

For a finite set of formulas $\Gamma \subseteq \mathsf{Fm}$, we say that a propositional variable X is *fresh for* Γ if $X \notin \bigcup_{F \in \Gamma} \text{PV}(F)$.

Lemma 10.21. *For every finite model* $\mathcal{M} = (\mathsf{v}, \mathcal{B})$ *and a formula* F*, there exists a propositional variable* $X_{F,\mathcal{B}}$ *that is fresh for* $\mathsf{V} \cup \{F\} \cup \text{subf}(\mathcal{B})$.

10. Atomic model semantics for JN$_V$

Proof. Since $\mathcal{M}$ is finite, so is $\mathcal{B}$ and therefore subf($\mathcal{B}$). V is finite. From this, the set

$$\bigcup_{P \in V} \mathrm{PV}(P) \cup \mathrm{PV}(F) \cup \bigcup_{H \in \mathrm{subf}(\mathcal{B})} \mathrm{PV}(H)$$

is finite as a finite union of finite sets, and there are infinitely many propositional variables in its complement since Prop is countable.

To be specific, take the smallest (in the sense of a fixed ordering of propositional variables) variable in the complement of that set. By definition, it is fresh for $V \cup \{F\} \cup \mathrm{subf}(\mathcal{B})$. □

10.2. Atomization

For the atomization procedure, we assume that we have an intermediate CS-model that satisfies a given formula F. We will also assume a locally finite CS.

As a starting point for the construction of the appropriate atomic model, we need a model that is "between" models from Theorems 9.10 and 9.11.

Lemma 10.22 (Starting finite model). *For every locally finite CS, intermediate CS-model $\mathcal{M}$ and formula F, the model*

$$\mathcal{M}' := (\mathsf{v} \cap \mathsf{v}_F, \mathrm{Fin}(\mathcal{B}, F) \cup \mathsf{CS}_F)^{\mathsf{CS}}$$

has the following properties:

(a) $\mathcal{M}'$ is a finite model.

(b) $\mathcal{M}'$ agrees with $\mathcal{M}$ on relevant propositional valuation for F.

(c) $\mathcal{M}'$ agrees with $\mathcal{M}$ on relevant evidence for F.

(d) $\mathcal{M}' \Vdash F \Leftrightarrow \mathcal{M} \Vdash F$.

(e) $\mathcal{M}'$ is an intermediate model.

(f) $\mathcal{M}'$ is a CS-closed model.

(g) $\mathcal{M}'$ contains CS_F.

Proof. For the purpose of the proof, we define the following bases:

$$\mathcal{B}_1 := \text{Fin}(\mathcal{B}, F)$$
$$\mathcal{B}_2 := \mathsf{CS}_F = \{(c, G) \mid (c, G) \in \mathsf{CS}, c \in \text{sub}(\mathcal{E}_F)\}$$
$$\mathcal{B}_3 := \{(c, G) \mid (c, G) \in \mathsf{CS}, c \in \text{sub}(\mathcal{B}_1 \cup \mathcal{B}_2)\}$$

Using this notation, $\mathcal{M}' = (\mathsf{v} \cap \mathsf{v}_F, \mathcal{B}_1 \cup \mathcal{B}_2)^{\mathsf{CS}} = (\mathsf{v} \cap \mathsf{v}_F, \mathcal{B}_1 \cup \mathcal{B}_2 \cup \mathcal{B}_3)$.

(a) $\mathcal{M}'$ *is a finite model.*

By Theorem 9.10, $(\mathsf{v} \cap \mathsf{v}_F, \text{Fin}(\mathcal{B}, F))$ is finite; hence, $\mathcal{B}_1 = \text{Fin}(\mathcal{B}, F)$ is finite.

For any formula F, $\text{sub}(\mathcal{E}_F)$ is finite, and for any finite basis $\tilde{\mathcal{B}}$ its set of subterms $\text{sub}(\tilde{\mathcal{B}})$ is finite as well.

For a locally finite CS, for every constant c, $\{(c, G) \mid (c, G) \in \mathsf{CS}\}$ is finite.

Therefore, $\mathcal{B}_2 = \bigcup_{c \in \text{sub}(\mathcal{E}_F)} \{(c, G) \mid (c, G) \in \mathsf{CS}\}$ is finite as a finite union of finite sets.

From this, $\mathcal{B}_1 \cup \mathcal{B}_2$ is finite, and consequently $\mathcal{B}_3$ is finite as well as a finite union of finite sets:

$$\mathcal{B}_3 = \bigcup_{c \in \text{sub}(\mathcal{B}_1 \cup \mathcal{B}_2)} \{(c, G) \mid (c, G) \in \mathsf{CS}\}$$

From this we can conclude that the model $\mathcal{M}' = (\mathsf{v} \cap \mathsf{v}_F, \mathcal{B}_1 \cup \mathcal{B}_2 \cup \mathcal{B}_3)$ is finite.

(b) $\mathcal{M}'$ *agrees with $\mathcal{M}$ on relevant propositional valuation for F.*

Shown in Theorem 9.10.

(c) $\mathcal{M}'$ *agrees with $\mathcal{M}$ on relevant evidence for F.*

The proof follows the same reasoning as in Theorem 9.10, with a minor addition in the proof of sub-claim $\mathcal{B}' \subseteq \mathcal{B}$.

Theorem 9.10 shows that $\mathcal{B}_1 = \text{Fin}(\mathcal{B}, F) \subseteq \mathcal{B}$.

Since $\mathcal{M}$ is a CS-model, $\mathsf{CS} \subseteq \mathcal{B}$ as well, and by construction $\mathcal{B}_2 \subseteq \mathsf{CS}$ and $\mathcal{B}_3 \subseteq \mathsf{CS}$.

Therefore, $\mathcal{B}' = \mathcal{B}_1 \cup \mathcal{B}_2 \cup \mathcal{B}_3 \subseteq \mathcal{B}$.

The rest of the argument is unchanged.

(d) $\mathcal{M}' \Vdash F \Leftrightarrow \mathcal{M} \Vdash F$.

As before, follows from two previous claims and Lemma 9.5.

(e) $\mathcal{M}'$ *is an intermediate model.*

Same as in Theorem 9.10.

(f) $\mathcal{M}'$ *is a* CS-*closed model.*

$\mathcal{M}'$ is the CS-closure of a model $(\mathsf{v} \cap \mathsf{v}_F, \mathrm{Fin}(\mathcal{B}, F) \cup \mathsf{CS}_F)$. By Lemma 10.5, it is CS-closed.

(g) $\mathcal{M}'$ *contains* CS_F.

By definition of CS-closure, model $\mathcal{M}'$ has the following basis:

$$\mathcal{B}' = \mathrm{Fin}(\mathcal{B}, F) \cup \mathsf{CS}_F \\ \cup \{(c, G) \mid (c, G) \in \mathsf{CS}, c \in \mathrm{sub}(\mathrm{Fin}(\mathcal{B}, F) \cup \mathsf{CS}_F)\}$$

Therefore, $\mathsf{CS}_F \subseteq \mathcal{B}'$ and by definition we conclude that $\mathcal{M}'$ contains CS_F.

□

Remark 10.23. Every term t can be uniquely represented using the form $t = \tilde{t} \cdot n_{P_k} \cdot \ldots \cdot n_{P_1}$ (possibly with $k = 0$, i.e. $t = \tilde{t}$), where $\tilde{t} \neq s \cdot n_Q$ for any s, Q.

As a reminder, term application is left-associative.

The following definition introduces, for a formula F and a model $\mathcal{M}$, a new model that agrees with $\mathcal{M}$ on truth of F but has less complexity.

Definition 10.24 (F-preserving atomization)**.** Take a finite intermediate model $\mathcal{M} = (\mathsf{v}, \mathcal{B})$ with $(t \cdot s, G) \in \mathcal{B}$ for some t, s, G.

By Remark 10.23, t and s can, without loss of generality, be uniquely represented as $\tilde{t} \cdot n_{P_k} \cdot \ldots \cdot n_{P_1}$ and $\tilde{s} \cdot n_{Q_l} \cdot \ldots \cdot n_{Q_1}$ such that neither $\tilde{t}$ nor $\tilde{s}$ can be further decomposed into $q \cdot n_R$ for any q, R.

By Lemma 10.21, there exists a variable $X := X_{F,\mathcal{B}}$ that is fresh for $(\mathsf{V} \cup \{F\} \cup \mathrm{subf}(\mathcal{B}))$.

10.2. Atomization

We define the *F-preserving atomization of* $(t \cdot s, G)$ *in* $\mathcal{M}$:

$$\mathrm{Atom}_F(\mathcal{M}; (t \cdot s, G)) := (\mathsf{v}, \mathcal{B}'), \text{ where}$$

$$\mathcal{B}' := (\mathcal{B} \setminus \{(t \cdot s, G)\}) \cup \left\{ \begin{array}{l} (\tilde{t}, P_k \to \ldots \to P_1 \to (X \to G)), \\ (\tilde{s}, Q_l \to \ldots \to Q_1 \to X) \end{array} \right\}$$

See Figure 2 for an illustration of the resulting applications.
Additionally, define a set of evidence pairs

$$\begin{aligned}
\mathcal{E}_X(t \cdot s, G) := \{ & \\
& (\tilde{s}, Q_l \to \ldots \to Q_1 \to X), \\
& (\tilde{s} \cdot n_{Q_l}, Q_{l-1} \to \ldots \to Q_1 \to X), \\
& \ldots, \\
& (\tilde{s} \cdot n_{Q_l} \cdot \ldots \cdot n_{Q_1}, X), \\
& (\tilde{t}, P_k \to \ldots \to P_1 \to (X \to G)), \\
& (\tilde{t} \cdot n_{P_k}, P_{k-1} \to \ldots \to P_1 \to (X \to G)), \\
& \ldots, \\
& (\tilde{t} \cdot n_{P_k} \cdot \ldots \cdot n_{P_1}, X \to G) \\
\}
\end{aligned}$$

All of these evidence pairs contain X as a subformula.
When the context is clear, we will simply refer to it as $\mathcal{E}_X$.

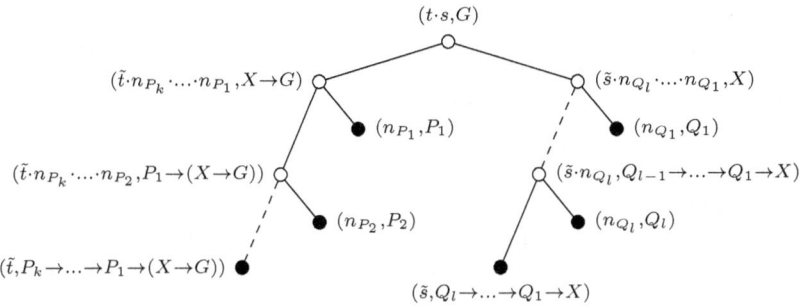

Figure 2: Atomization of $(t \cdot s, G)$

10. Atomic model semantics for JN$_V$

The idea of atomization of $(t \cdot s, G)$ is to create a new basis of lesser complexity that will still contain $(t \cdot s, G)$ and the rest of $\mathcal{B}$ in $\mathcal{E}(\mathcal{B})$. By choice of X, the new model will agree on relevant evidence for F, therefore "preserving" it.

With the following lemma, we want to show that $\mathcal{E}(\mathcal{B}')$ consists of $\mathcal{E}(\mathcal{B})$ and evidence pairs that contain X as subformulas (and therefore not relevant to F, since X was chosen fresh).

Lemma 10.25. *For a finite intermediate model* $\mathcal{M} = (\mathsf{v}, \mathcal{B})$, $(t \cdot s, G) \in \mathcal{B}$ *and a formula* F, *take* $(\mathsf{v}, \mathcal{B}') := \mathrm{Atom}_F(\mathcal{M}; (t \cdot s, G))$. *Then,*

$$(q, H) \in \mathcal{E}(\mathcal{B}') \quad \Rightarrow \quad \begin{cases} (q, H) \in \mathcal{E}(\mathcal{B}), & \text{or} \\ (q, H) = (\Sigma[r], H) \text{ for some } & \Sigma \in \mathcal{S}(\mathcal{B})^*, \\ & (r, H) \in \mathcal{E}_X. \end{cases}$$

Note that $\Sigma \in \mathcal{S}(\mathcal{B})^*$ *and not* $\mathcal{S}(\mathcal{B}')^*$.

Proof. We shall call the two cases of the claim as case (A) or case (B) respectively further in this proof.

Assume $(q, H) \in \mathcal{E}(\mathcal{B}')$ Proof by induction on the buildup of $\mathcal{E}(\mathcal{B}')$:
Base: $(q, H) \in \mathcal{B}'$. By definition of $\mathcal{B}'$, there are two cases:

(a) $(q, H) \in \mathcal{B} \setminus \{(t \cdot s, G)\}$.

Since $\mathcal{B} \setminus \{(t \cdot s, G)\} \subseteq \mathcal{B} \subseteq \mathcal{E}(\mathcal{B})$ by Lemma 7.16, the claim is valid by case (A).

(b) $(q, H) = (\tilde{t}, P_k \to \ldots \to P_1 \to (X \to G))$.

Then $(\lambda[q], H) = (q, H) \in \mathcal{E}_X$ and the claim is valid by case (B).

(c) $(q, H) = (\tilde{s}, Q_l \to \ldots \to Q_1 \to X)$.

Same as the above.

App: $(q \cdot r, H) \in \mathcal{E}(\mathcal{B}')$, and $\{(q, E \to H), (r, E)\} \subseteq \mathcal{E}(\mathcal{B}')$.

Induction hypothesis applies to $(q, E \to H)$ and (r, E); therefore, there are 4 distinct cases based on whether case (A) or (B) applies for them.

(a) Case (A, A):

$\{(q, E \to H), (r, E)\} \subseteq \mathcal{E}(\mathcal{B})$.

10.2. Atomization

Then, by Lemma 7.16, $(q \cdot r, H) \in \mathcal{E}(\mathcal{B})$, which shows the case (A) of the claim.

(b) Case (A, B):

$(q, E \to H) \in \mathcal{E}(\mathcal{B})$, $r = \Sigma[\tilde{r}]$ for some $\Sigma \in \mathcal{S}(\mathcal{B})^*$, $(\tilde{r}, E) \in \mathcal{E}_X$.

This case is impossible: Since $(\tilde{r}, E) \in \mathcal{E}_X$, X must be a subformula of E, however $(q, E \to H) \in \mathcal{E}(\mathcal{B})$ and therefore X cannot be a subformula of $E \to H$, which contains all subformulas of E.

(c) Case (B, A):

$q = \Sigma[\tilde{q}]$ for some $\Sigma \in \mathcal{S}(\mathcal{B})^*$, $(\tilde{q}, E \to H) \in \mathcal{E}_X$ and $(r, E) \in \mathcal{E}(\mathcal{B})$.

From this we conclude that X is a subformula of $E \to H$ but not E.

That leaves the following possibilities for $(\tilde{q}, E \to H) \in \mathcal{E}_X$:

$$\tilde{q} = \tilde{s} \cdot n_{Q_l} \cdot \ldots \cdot n_{Q_{l-i+1}}, \qquad E = Q_{l-i},$$
$$H = Q_{l-i-1} \ldots \to Q_1 \to X \qquad 0 \leqslant i < l$$

or

$$\tilde{q} = \tilde{t} \cdot n_{P_k} \cdot \ldots \cdot n_{P_{k-j+1}}, \qquad E = P_{k-j},$$
$$H = P_{k-j-1} \ldots \to P_1 \to (X \to G) \qquad 0 \leqslant j < k$$

We will prove the claim for the first form, and the proof for the second form is analogous.

Since $(r, E) = (r, Q_{l-i}) \in \mathcal{E}(\mathcal{B})$, we have $\langle 0, Q_{l-i}, r \rangle \in \mathcal{S}(\mathcal{B})$.

Note that the term r can be represented as $\langle 0, Q_{l-i}, r \rangle [n_{Q_{l-i}}]$.

Define $\Sigma' := \text{Comb}(\Sigma, \tilde{s} \cdot n_{Q_l} \cdot \ldots \cdot n_{Q_{l-i+1}}, \langle 0, Q_{l-i}, r \rangle)$.

With that, by Lemma 10.18,

$$(q \cdot r, H) = (\Sigma[\tilde{s} \cdot n_{Q_l} \cdot \ldots \cdot n_{Q_{l-i+1}}] \cdot \langle 0, Q_{l-i}, r \rangle [n_{Q_{l-i}}], H)$$
$$= (\Sigma'[\tilde{s} \cdot n_{Q_l} \cdot \ldots \cdot n_{Q_{l-i+1}} \cdot n_{Q_{l-i}}], H)$$
$$= \left(\Sigma'[\underbrace{\tilde{s} \cdot n_{Q_l} \cdot \ldots \cdot n_{Q_{l-i+1}} \cdot n_{Q_{l-i}}}_{v}], Q_{l-i-1} \ldots \to Q_1 \to X \right)$$

This means that $(q \cdot r, H) = (\Sigma'[v], H)$ for $\Sigma' \in \mathcal{S}(\mathcal{B})^*$ and

$$(v, H) = (\tilde{s} \cdot n_{Q_l} \cdot \ldots \cdot n_{Q_{l-i+1}} \cdot n_{Q_{l-i}}, Q_{l-i-1} \ldots \to Q_1 \to X) \in \mathcal{E}_X,$$

and therefore the claim is valid by case (B).

(d) Case (B, B):
$q = \Sigma_1[\tilde{q}]$, $r = \Sigma_2[\tilde{r}]$ for some substitutions $\Sigma_1, \Sigma_2 \in \mathcal{S}(\mathcal{B})^*$, and $\{(\tilde{q}, E \to H), (\tilde{r}, E)\} \subseteq \mathcal{E}_X$.

There is exactly one possibility for both of those evidence pairs being in $\mathcal{E}_X$, specifically

$$(\tilde{q}, E \to H) = (\tilde{t} \cdot n_{P_k} \cdot \ldots \cdot n_{P_1}, X \to G) = (t, X \to G),$$
$$(\tilde{r}, E) = (\tilde{s} \cdot n_{Q_l} \cdot \ldots \cdot n_{Q_1}, X) = (s, X)$$

Therefore, $(\tilde{q} \cdot \tilde{r}, H) = (t \cdot s, G) \in \mathcal{E}(\mathcal{B})$.
Take $\Sigma := \text{Comb}(\Sigma_1, \tilde{q}, \Sigma_2)$.
By Lemma 10.18, $\Sigma \in \mathcal{S}(\mathcal{B})^*$ and

$$(q \cdot r, H) = (\Sigma_1[\tilde{q}] \cdot \Sigma_2[\tilde{r}], H) = (\Sigma(t \cdot s), G)$$

Since $(t \cdot s, G) \in \mathcal{E}(\mathcal{B})$, by Lemma 10.12 we have $(q \cdot r, H) \in \mathcal{E}(\mathcal{B})$ and the claim is valid by case (A).

Substitution: $(\sigma q, H) \in \mathcal{E}(\mathcal{B}')$ for $\sigma = \langle i, R, r \rangle \in \mathcal{S}(\mathcal{B}')^*$, and (IH) applies for $\{(q, H), (r, R)\} \subseteq \mathcal{E}(\mathcal{B}')$.
Consider that $R \in V$ and $X \notin V$ by choice of X.
From this, $X \notin \text{subf}(R) = \{R\}$. Therefore, $(r, R) \notin \mathcal{E}_X$.
By (IH), $(r, R) \in \mathcal{E}(\mathcal{B})$ which means that $\sigma \in \mathcal{S}(\mathcal{B})$.
Two cases depending on which case of (IH) applies for (q, H):

(a) Case (A): $(q, H) \in \mathcal{E}(\mathcal{B})$.
In that case, $(\sigma q, H) = (\sigma[q], H) \in \mathcal{E}(\mathcal{B})$ by Lemma 10.12 and the claim is valid by case (A).

(b) Case (B): $(q, H) = (\Sigma[\tilde{q}], H)$ for some $\Sigma \in \mathcal{S}(\mathcal{B})^*$, $(\tilde{q}, H) \in \mathcal{E}_X$.
Then $(\sigma q, H) = (\sigma[\Sigma[\tilde{q}]], H) = ((\sigma \circ \Sigma)[\tilde{q}], H)$.

10.2. Atomization

By definition of $\mathcal{S}(\mathcal{B})^*$ and the fact that $\sigma \in \mathcal{S}(\mathcal{B})$, $\sigma \circ \Sigma \in \mathcal{S}(\mathcal{B})^*$ and therefore the claim is valid by case (B).

$\square$

Next, we will show that F-preserving atomization step has all the required properties to reduce complexity while maintaining the truth value of F.

Lemma 10.26. *For a finite intermediate model $\mathcal{M} = (\mathsf{v}, \mathcal{B})$, evidence pair $(t \cdot s, G) \in \mathcal{B}$, and a formula F, $(\mathsf{v}, \mathcal{B}') := \mathrm{Atom}_F(\mathcal{M}; (t \cdot s, G))$ has the following properties:*

(a) $\mathrm{Atom}_F(\mathcal{M}; (t \cdot s, G))$ is finite.

(b) $\mathrm{Atom}_F(\mathcal{M}; (t \cdot s, G))$ is intermediate.

(c) $\mathrm{Atom}_F(\mathcal{M}; (t \cdot s, G))$ agrees with $\mathcal{M}$ on relevant propositional valuation for F.

(d) $\mathrm{Atom}_F(\mathcal{M}; (t \cdot s, G))$ agrees with $\mathcal{M}$ on relevant evidence for F.

(e) $\mathrm{Atom}_F(\mathcal{M}; (t \cdot s, G))$ is CS-closed if $\mathcal{M}$ is CS-closed.

(f) $\mathrm{Atom}_F(\mathcal{M}; (t \cdot s, G))$ contains CS_F if $\mathcal{M}$ contains CS_F.

(g) $\mathrm{cmp}(\mathrm{Atom}_F(\mathcal{M}; (t \cdot s, G))) < \mathrm{cmp}(\mathcal{M})$.

Proof. (a) $\mathcal{M}$ is finite, or equivalently both $\mathcal{B}$ and v are finite.

$\mathcal{B}'$, by definition, has exactly 1 more element than $\mathcal{B}$ and is therefore finite as well.

Therefore, $\mathrm{Atom}_F(\mathcal{M}; (t \cdot s, G)) = (\mathsf{v}, \mathcal{B}')$ is finite.

(b) We need to show (R.1) — (R.3) for $\mathcal{B}'$. Note that $\mathcal{M}$ is intermediate, and therefore (R.1) — (R.3) holds for $\mathcal{B}$.

(R.1, $\Leftarrow$) We want to show that $(n_P, P) \in \mathcal{B}'$ for $P \in \mathsf{V}$.

Since $\mathcal{M}$ is intermediate, $(n_P, P) \in \mathcal{B}$.

Since n_P is atomic, $(n_P, P) \in (\mathcal{B} \setminus \{(t \cdot s, G)\}) \subseteq \mathcal{B}'$.

(R.1, $\Rightarrow$) Assume $(n_P, H) \in \mathcal{B}'$ with $P \in \mathsf{V}$.

There are three cases:

- $(n_P, H) \in (\mathcal{B} \setminus \{(t \cdot s, G)\}) \subseteq \mathcal{B}$.

  Since $\mathcal{M}$ is intermediate, by (R.1) for $\mathcal{B}$ we have $H = P$.

- $(n_P, H) = (\tilde{t}, P_k \to \ldots \to P_1 \to (X \to G))$.

  This case is impossible: if $\tilde{t} = n_P$, then $\tilde{t} \cdot n_{P_k}$ (or, in case $k = 0$, $\tilde{t} \cdot s$) is a subterm of $t \cdot s$ for $(t \cdot s, F) \in \mathcal{B}$, which is impossible by (R.2) for $\mathcal{B}$.

- $(n_P, H) = (\tilde{s}, Q_l \to \ldots \to Q_1 \to X)$.

  This case is again impossible. If $l \neq 0$, then the argument is as above with $\tilde{s} \cdot n_{Q_l}$.

  If $l = 0$, then $(t \cdot s, F) = (t \cdot n_P, F) \in \mathcal{B}$, which contradicts (R.3) for $\mathcal{B}$.

(R.2) Assume that (R.2) does not hold and $n_P \cdot r$ is a subterm of q for some $(q, H) \in \mathcal{B}'$.

There are three cases:

- $(q, H) \in (\mathcal{B} \setminus \{(t \cdot s, G)\}) \subseteq \mathcal{B}$.

  Since $\mathcal{M}$ is intermediate, (R.2) applies for it.

  Consequently, $n_P \cdot r \in \mathrm{sub}(q)$ is impossible.

- $(q, H) = (\tilde{t}, P_k \to \ldots \to P_1 \to (X \to G))$.

  By choice of $\tilde{t}$, $\tilde{t} \in \mathrm{sub}(t)$.

  Therefore, $\mathrm{sub}(q) \subseteq \mathrm{sub}(t) \subseteq \mathrm{sub}(t \cdot s)$ and $(t \cdot s, G) \in \mathcal{B}$.

  (R.2) applies for $\mathcal{M}$, therefore $n_P \cdot r \in \mathrm{sub}(t \cdot s)$ is impossible.

- $(q, H) = (\tilde{s}, Q_l \to \ldots \to Q_1 \to X)$.

  Analogous to the above case.

Therefore, we obtain a contradiction, which proves (R.2).

(R.3) Assume that (R.3) does not hold and $(q \cdot n_P, H) \in \mathcal{B}'$ for some q, P, H.

There are three cases:

- $(q \cdot n_P, H) \in (\mathcal{B} \setminus \{(t \cdot s, G)\}) \subseteq \mathcal{B}$.

  Therefore, $(q \cdot n_P, H) \in \mathcal{B}$, which is impossible since $\mathcal{M}$ is intermediate and (R.3) applies for it.

- $(q \cdot n_P, H) = (\tilde{t}, P_k \to \ldots \to P_1 \to (X \to G))$.

  By choice of $\tilde{t}$ according to Remark 10.23, $\tilde{t} = q \cdot n_P$ is impossible.

- $(q \cdot n_P, H) = (\tilde{s}, Q_l \to \ldots \to Q_1 \to X)$.

  Analogous to the above case.

Therefore, we obtain a contradiction, which proves (R.3).

(c) The claim is trivial, since $\mathcal{M}$ and $\mathrm{Atom}_F(\mathcal{M}; (t \cdot s, G))$ have the same propositional valuation v, and therefore agree on relevant propositional valuation for any formula.

(d) We need to show that $\mathcal{E}(\mathcal{B}) \cap \mathcal{E}_F = \mathcal{E}(\mathcal{B}') \cap \mathcal{E}_F$.

Recall the definition of $\mathcal{E}_X$ from Definition 10.24.

As a first step, we show that $\mathcal{E}_X \subseteq \mathcal{E}(\mathcal{B}')$ by applying Lemma 7.16 multiple times:

$$(\tilde{s}, Q_l \to \ldots \to Q_1 \to X) \in \mathcal{B}' \Rightarrow (\tilde{s}, Q_l \to \ldots \to Q_1 \to X) \in \mathcal{E}(\mathcal{B}')$$
$$(n_{Q_l}, Q_l) \in (\mathcal{B} \setminus \{(t \cdot s, G)\}) \subseteq \mathcal{B}' \Rightarrow (n_{Q_l}, Q_l) \in \mathcal{E}(\mathcal{B}')$$
$$\text{(conclusions above)} \Rightarrow (\tilde{s} \cdot n_{Q_l}, Q_{l-1} \to \ldots \to Q_1 \to X) \in \mathcal{E}(\mathcal{B}')$$
$$\ldots$$
$$\text{(conclusions above)} \Rightarrow (\tilde{s} \cdot n_{Q_l} \cdot \ldots \cdot n_{Q_1}, X) = (s, X) \in \mathcal{E}(\mathcal{B}')$$

And similarly starting from $(\tilde{t}, P_k \to \ldots \to P_1 \to (X \to G))$.

Note that from this we have $\{(t, X \to G), (s, X)\} \subseteq \mathcal{E}_X \subseteq \mathcal{E}(\mathcal{B}')$, and by applying Lemma 7.16 again we get $(t \cdot s, G) \in \mathcal{E}(\mathcal{B}')$.

By construction of $\mathcal{B}'$ and Lemma 7.16, we have

$$\mathcal{B} \setminus \{(t \cdot s, G)\} \subseteq \mathcal{B}' \subseteq \mathcal{E}(\mathcal{B}')$$

Combining the last two conclusions, we obtain $\mathcal{B} \subseteq \mathcal{E}(\mathcal{B}')$.

By Lemma 9.6, $\mathcal{E}(\mathcal{B}) \subseteq \mathcal{E}(\mathcal{E}(\mathcal{B}')) = \mathcal{E}(\mathcal{B}')$.

From this we immediately obtain $\mathcal{E}(\mathcal{B}) \cap \mathcal{E}_F \subseteq \mathcal{E}(\mathcal{B}') \cap \mathcal{E}_F$.

10. Atomic model semantics for JN$_V$

To show the inverse inclusion, $\mathcal{E}(\mathcal{B}') \cap \mathcal{E}_F \subseteq \mathcal{E}(\mathcal{B}) \cap \mathcal{E}_F$, assume that $(q, H) \in \mathcal{E}(\mathcal{B}') \cap \mathcal{E}_F$.

By Lemma 10.25, we have two cases for $(q, H) \in \mathcal{E}(\mathcal{B}')$:

a) $(q, H) \in \mathcal{E}(\mathcal{B})$.

   Therefore, $(q, H) \in \mathcal{E}(\mathcal{B}) \cap \mathcal{E}_F$ as required.

b) $(q, H) = (\Sigma[r], H)$ for some $\Sigma \in \mathcal{S}(\mathcal{B})^*, (r, H) \in \mathcal{E}_X$.

   Since $(r, H) \in \mathcal{E}_X$, X occurs in H. By choice of X, X does not occur in $\text{subf}(\mathcal{E}_F)$. Therefore, $(q, H) \notin \mathcal{E}_F$ and this case does not apply.

(e) To prove CS-closedness we must show the following:

$$\left(\bigcup_{(q,H) \in \mathcal{B}'} \{(c, E) \mid (c, E) \in \text{CS}, c \in \text{sub}(q)\} \right) \subseteq \mathcal{B}'$$

Observe that all terms occurring in $\mathcal{E}_X$ are subterms of $t \cdot s$. Therefore, all constants occurring in $\mathcal{E}_X$ already occur in $\mathcal{B}$.

There are two cases for $(q, H) \in \mathcal{B}'$:

a) $(q, H) \in \mathcal{B} \setminus \{(t \cdot s, G)\}$.

   Since $\mathcal{M}$ is CS-closed,

$$\left(\bigcup_{(q,H) \in \mathcal{B}} \{(c, E) \mid (c, E) \in \text{CS}, c \in \text{sub}(q)\} \right) \subseteq \mathcal{B}$$

   and in particular $\{(c, E) \mid (c, E) \in \text{CS}, c \in \text{sub}(q)\} \subseteq \mathcal{B}$ for (q, H).

   Since $(t \cdot s, G) \neq (c, E)$ for any c, E, we conclude

$$\{(c, E) \mid (c, E) \in \text{CS}, c \in \text{sub}(q)\} \subseteq \mathcal{B} \setminus \{(t \cdot s, G)\} \subseteq \mathcal{B}'$$

b) $(q, H) \in \left\{ \begin{array}{l} (\tilde{t}, P_k \to \ldots \to P_1 \to (X \to G)), \\ (\tilde{s}, Q_l \to \ldots \to Q_1 \to X) \end{array} \right\}$

110

10.2. Atomization

In either case, since $q \in \mathrm{sub}(t \cdot s)$,

$$\{(c, E) \mid (c, E) \in \mathsf{CS}, c \in \mathrm{sub}(q)\}$$
$$\subseteq \{(c, E) \mid (c, E) \in \mathsf{CS}, c \in \mathrm{sub}(t \cdot s)\}$$

And since $\mathcal{M}$ is CS-closed and $(t \cdot s, G) \in \mathcal{B}$, we have

$$\{(c, E) \mid (c, E) \in \mathsf{CS}, c \in \mathrm{sub}(q)\} \subseteq \mathcal{B}$$

Finally, by the same argument as in the previous case

$$\{(c, E) \mid (c, E) \in \mathsf{CS}, c \in \mathrm{sub}(q)\} \subseteq \mathcal{B} \setminus \{(t \cdot s, G)\} \subseteq \mathcal{B}'$$

This shows that $\{(c, E) \mid (c, E) \in \mathsf{CS}, c \in \mathrm{sub}(q)\} \subseteq \mathcal{B}'$ for any $(q, H) \in \mathcal{B}'$, or equivalently

$$\left(\bigcup_{(q,H) \in \mathcal{B}'} \{(c, E) \mid (c, E) \in \mathsf{CS}, c \in \mathrm{sub}(q)\} \right) \subseteq \mathcal{B}'.$$

(f) It's easy to see that $\mathcal{B} \cap \mathsf{CS} \subseteq \mathcal{B}' \cap \mathsf{CS}$: since $(t \cdot s, G) \notin \mathsf{CS}$ we have

$$\mathcal{B} \cap \mathsf{CS} = (\mathcal{B} \setminus \{(t \cdot s, G)\}) \cap \mathsf{CS} \subseteq \mathcal{B}' \cap \mathsf{CS}$$

Therefore, if $\mathsf{CS}_F \subseteq \mathcal{B}$, and considering that $\mathsf{CS}_F \subseteq \mathsf{CS}$, we have

$$\mathsf{CS}_F = \mathsf{CS}_F \cap \mathsf{CS} \subseteq \mathcal{B} \cap \mathsf{CS} \subseteq \mathcal{B}' \cap \mathsf{CS} \subseteq \mathcal{B}'$$

(g) By definition, $\mathrm{cmp}(\mathrm{Atom}_F(\mathcal{M}; (t \cdot s, G))) = \mathrm{cmp}(\mathcal{B}')$.

In general, we have $\mathrm{cmp}(\mathcal{B}_1 \cup \mathcal{B}_2) \leqslant \mathrm{cmp}(\mathcal{B}_1) + \mathrm{cmp}(\mathcal{B}_2)$ and the inequality becomes equality in case of a disjoint union.

Considering the above, definition of $\mathcal{B}'$ and $\mathrm{cmp}(\{(q, H)\}) = \mathrm{cmp}(q)$, we have:

$$\mathrm{cmp}(\mathcal{B}') \leqslant \mathrm{cmp}(\mathcal{B} \setminus \{(t \cdot s, G)\}) + \mathrm{cmp}(\tilde{t}) + \mathrm{cmp}(\tilde{s})$$

Since $(t \cdot s, G) \in \mathcal{B}$ and the union $(\mathcal{B} \setminus \{(t \cdot s, G)\}) \cup \{(t \cdot s, G)\}$ is disjoint, we have

$$\mathrm{cmp}(\mathcal{B}) = \mathrm{cmp}((\mathcal{B} \setminus \{(t \cdot s, G)\}) \cup \{(t \cdot s, G)\})$$
$$= \mathrm{cmp}(\mathcal{B} \setminus \{(t \cdot s, G)\}) + \mathrm{cmp}(t \cdot s)$$

Therefore,

$$\mathrm{cmp}(\mathrm{Atom}_F(\mathcal{M};(t \cdot s, G))) \leqslant \mathrm{cmp}(\mathcal{B}) - \mathrm{cmp}(t \cdot s) + \mathrm{cmp}(\tilde{t}) + \mathrm{cmp}(\tilde{s})$$
$$= \mathrm{cmp}(\mathcal{B}) - (\mathrm{cmp}(t) + \mathrm{cmp}(s) + 1) + \mathrm{cmp}(\tilde{t}) + \mathrm{cmp}(\tilde{s})$$
$$= \mathrm{cmp}(\mathcal{B}) - 1 - (\mathrm{cmp}(t) - \mathrm{cmp}(\tilde{t})) - (\mathrm{cmp}(s) - \mathrm{cmp}(\tilde{s}))$$

Note that by choice of $\tilde{t}$, $\mathrm{cmp}(t) \geqslant \mathrm{cmp}(\tilde{t})$ and similarly for s. This shows the claim:

$$\mathrm{cmp}(\mathrm{Atom}_F(\mathcal{M};(t \cdot s, G)))$$
$$\leqslant \mathrm{cmp}(\mathcal{B}) - 1 - (\mathrm{cmp}(t) - \mathrm{cmp}(\tilde{t})) - (\mathrm{cmp}(s) - \mathrm{cmp}(\tilde{s}))$$
$$\leqslant \mathrm{cmp}(\mathcal{B}) - 1$$
$$= \mathrm{cmp}(\mathcal{M}) - 1 < \mathrm{cmp}(\mathcal{M})$$

□

Lemma 10.27 (Atomization step). *For any finite intermediate* CS-*closed model* $\mathcal{M} = (\mathsf{v}, \mathcal{B})$ *with complexity* $\mathrm{cmp}(\mathcal{M}) > 0$ *and a formula F such that $\mathcal{M} \Vdash F$ and $\mathcal{M}$ contains* CS_F, *there exists a finite intermediate* CS-*closed model* $\mathcal{M}'$ *containing* CS_F *with complexity* $\mathrm{cmp}(\mathcal{M}') < \mathrm{cmp}(\mathcal{M})$ *and* $\mathcal{M}' \Vdash F$.

Proof. Since $\mathrm{cmp}(\mathcal{M}) > 0$, that means there exists $(t \cdot s, G) \in \mathcal{B}$.

Since $\mathcal{M}$ is also a finite intermediate model, it fulfills the requirements for defining $\mathrm{Atom}_F(\mathcal{M};(t \cdot s, G))$.

Take $\mathcal{M}' := \mathrm{Atom}_F(\mathcal{M};(t \cdot s, G))$.

By Lemma 10.26, $\mathcal{M}'$ is CS-closed, finite, intermediate, fits the complexity bound and contains CS_F.

Since $\mathcal{M}$ and $\mathcal{M}'$ agree on relevant propositional valuation and evidence for F, by Lemma 9.5 we have $\mathcal{M}' \Vdash F$. □

By iterating the atomization step, we can reduce the complexity of the model to zero, yielding an atomic model.

10.2. Atomization

Lemma 10.28 (Atomization). *For any finite intermediate CS-closed model $\mathcal{M} = (\mathsf{v}, \mathcal{B})$ and a formula F such that $\mathcal{M} \Vdash F$ and $\mathcal{M}$ contains CS_F, there exists a finite atomic CS-closed model $\mathcal{M}_a$ containing CS_F such that $\mathcal{M}_a \Vdash F$.*

Proof. The complexity $\mathrm{cmp}(\mathcal{M})$ is finite since the model is finite.
Proof by induction on $\mathrm{cmp}(\mathcal{M})$:

- Base: $\mathrm{cmp}(\mathcal{M}) = 0$.

  In this case $\mathcal{M}$ is already atomic and one can take $\mathcal{M}_a := \mathcal{M}$.

- Step: $\mathrm{cmp}(\mathcal{M}) > 0$, and the induction hypothesis applies for models of lower complexity.

  Apply Lemma 10.27 to $\mathcal{M}$: we obtain $\mathcal{M}'$ with a strictly smaller complexity that fits the premises of the claim.

  Therefore, induction hypothesis applies to $\mathcal{M}'$ and we have finite atomic CS-closed model $\mathcal{M}'_a$ containing CS_F such that $\mathcal{M}'_a \Vdash F$.

  Take $\mathcal{M}_a := \mathcal{M}'_a$.

$\square$

However, the obtained atomic model may not yet be a CS-model. We intend to add the entire constant specification back into the basis, but we must prove that this will not create relevant evidence for F. The fact that we kept the model CS-closed will guarantee it, as any new evidence will have to contain new constants.

Lemma 10.29. *Given a CS-closed model $\mathcal{M} = (\mathsf{v}, \mathcal{B})$,*

$$(q, G) \in \mathcal{E}(\mathcal{B} \cup \mathsf{CS}) \quad \Rightarrow \quad \begin{cases} (q, G) \in \mathcal{E}(\mathcal{B}), \text{ or} \\ \exists c : c \in \mathrm{sub}(q) \setminus \mathrm{sub}(\mathcal{B}), c \in \mathrm{sub}(\mathsf{CS}) \end{cases}$$

Proof. We shall call the two cases of the claim as case (A) or case (B) respectively further in this proof.
Assume $(q, G) \in \mathcal{E}(\mathcal{B} \cup \mathsf{CS})$. Induction on the buildup of $\mathcal{E}(\mathcal{B} \cup \mathsf{CS})$:
<u>Base:</u> $(q, G) \in \mathcal{B} \cup \mathsf{CS}$.

- If $(q, G) \in \mathcal{B}$, we have $(q, G) \in \mathcal{E}(\mathcal{B})$ by Lemma 7.16 and the claim is shown by case (A).

- Suppose $(q, G) = (c, G) \in \mathsf{CS} \setminus \mathcal{B}$.

  If we assume $c \in \text{sub}(\mathcal{B})$, then by CS-closedness we have $(c, G) \in \mathcal{B}$, which contradicts $(c, G) \in \mathsf{CS} \setminus \mathcal{B}$.

  Therefore, $c \notin \text{sub}(\mathcal{B})$, while obviously $c \in \text{sub}(c)$. This shows the claim by case (B).

Application: $(t \cdot s, G) \in \mathcal{E}(\mathcal{B} \cup \mathsf{CS})$ since $(t, H \to G) \subseteq \mathcal{E}(\mathcal{B} \cup \mathsf{CS})$ and $(s, H) \subseteq \mathcal{E}(\mathcal{B} \cup \mathsf{CS})$.
Induction hypothesis applies for $(t, H \to G)$ and (s, H).

- If case (A) applies to both, $(t, H \to G) \in \mathcal{E}(\mathcal{B})$ and $(s, H) \in \mathcal{E}(\mathcal{B})$.

  Therefore, by Lemma 7.16, $(t \cdot s, G) \in \mathcal{E}(\mathcal{B})$ and claim is shown by case (A).

- Otherwise, there exists a constant c in either $(\text{sub}(t) \setminus \text{sub}(\mathcal{B}))$ or $(\text{sub}(s) \setminus \text{sub}(\mathcal{B}))$.

  Since $\text{sub}(t \cdot s) = \{t \cdot s\} \cup \text{sub}(s) \cup \text{sub}(t)$, we have $c \in \text{sub}(t \cdot s) \setminus \text{sub}(\mathcal{B})$ and claim is shown by case (B).

Substitution:
$(\sigma q, H) \in \mathcal{E}(\mathcal{B} \cup \mathsf{CS})$ for $\sigma = \langle i, P, r \rangle$, since $\{(q, H), (r, P)\} \subseteq \mathcal{E}(\mathcal{B} \cup \mathsf{CS})$.
Induction hypothesis applies to (q, H) and (r, P).
By Lemma 10.19, we have 2 sub-cases:

- $\sigma q = q$.

  If case (A) applies for (q, H), then $(\sigma q, H) = (q, H) \in \mathcal{E}(\mathcal{B})$ showing case (A) for this branch.

  Otherwise, case (B) applies for (q, H) and we have some constant c such that $c \in \text{sub}(q) \setminus \text{sub}(\mathcal{B})$.

  However, $\text{sub}(\sigma q) = \text{sub}(q)$ and therefore case (B) applies for this branch.

- For every constant c' such that $c' \in \text{sub}(q) \cup \text{sub}(r)$, $c' \in \text{sub}(\sigma q)$.

  If case (A) applies for both (q, H) and (r, P), we have $(q, H) \in \mathcal{E}(\mathcal{B})$ and $(r, P) \in \mathcal{E}(\mathcal{B})$.

  Therefore, by Lemma 7.16, $(\sigma q, H) \in \mathcal{E}(\mathcal{B})$ and claim is shown by case (A).

10.2. Atomization

Otherwise, case (B) applies for either (q, H) or (r, P), and therefore there exists a constant c that belongs to either $(\text{sub}(q) \setminus \text{sub}(\mathcal{B}))$ or $(\text{sub}(r) \setminus \text{sub}(\mathcal{B}))$, and $c \in \text{sub}(\text{CS})$.

From the above we can conclude that $c \in \text{sub}(q) \cup \text{sub}(r)$, which yields $c \in \text{sub}(\sigma q)$ and shows case (B) for this branch.

□

Note that the condition to contain CS_F in the model $\mathcal{M}$ is key for preserving truth of F when "adding back" the full constant specification CS.

If we did not guarantee this condition, adding back the constant specification may produce new relevant evidence for F, potentially making F false in $\mathcal{M}'$.

Lemma 10.30. *Given a formula F and a finite atomic CS-closed model $\mathcal{M} = (\mathsf{v}, \mathcal{B})$ containing CS_F such that $\mathcal{M} \Vdash F$, there exists an almost finite atomic CS-model $\mathcal{M}'$ such that $\mathcal{M}' \Vdash F$.*

Proof. Take $\mathcal{M}' := (\mathsf{v}, \mathcal{B} \cup \text{CS})$. By definition, $\mathcal{M}'$ is a CS-model. Since $\mathcal{B}$ is finite, by definition $\mathcal{M}'$ is almost finite.

$\mathcal{M}$ is atomic: for any $(t, G) \in \mathcal{B}$, t is atomic.

For any $(t, G) \in \text{CS}$, t must be a constant and therefore atomic.

From this, for any $(t, G) \in \mathcal{B} \cup \text{CS}$, t must be atomic, and therefore $\mathcal{M}'$ is atomic.

Finally, we want to show that $\mathcal{M}' \Vdash F$, by using Lemma 9.5. $\mathcal{M}$ and $\mathcal{M}'$ have the same propositional valuation v, and therefore agree on relevant propositional valuation for any formula. All that's left to show is that $\mathcal{M}$ and $\mathcal{M}'$ agree on relevant evidence for F.

Since $\mathcal{B} \subseteq \mathcal{B} \cup \text{CS}$, we have $\mathcal{E}(\mathcal{B}) \subseteq \mathcal{E}(\mathcal{B} \cup \text{CS})$ and therefore we have $\mathcal{E}(\mathcal{B}) \cap \mathcal{E}_F \subseteq \mathcal{E}(\mathcal{B} \cup \text{CS}) \cap \mathcal{E}_F$.

We need to show that $\mathcal{E}(\mathcal{B} \cup \text{CS}) \cap \mathcal{E}_F \subseteq \mathcal{E}(\mathcal{B}) \cap \mathcal{E}_F$.

Suppose $(t, G) \in \mathcal{E}(\mathcal{B} \cup \text{CS}) \cap \mathcal{E}_F$.

By Lemma 10.29, there are two cases from $(t, G) \in \mathcal{E}(\mathcal{B} \cup \text{CS})$:

(a) $(t, G) \in \mathcal{E}(\mathcal{B})$. In this case, we immediately obtain $(t, G) \in \mathcal{E}(\mathcal{B}) \cap \mathcal{E}_F$.

(b) There exists a constant c such that both $c \in \text{sub}(t) \setminus \text{sub}(\mathcal{B})$ and $c \in \text{sub}(\text{CS})$.

We assume that $(t, G) \in \mathcal{E}_F$. From this, $c \in \text{sub}(t) \subseteq \text{sub}(\mathcal{E}_F)$.

10. Atomic model semantics for JN$_V$

Since $c \in \mathrm{sub}(\mathsf{CS})$, there exists $(c', F') \in \mathsf{CS}$ such that

$$c \in \mathrm{sub}(c') = \{c'\}$$

Therefore, $c' = c$ and $(c, F) \in \mathsf{CS}$.
However, since $c \in \mathrm{sub}(\mathcal{E}_F)$, we conclude that $(c, F) \in \mathsf{CS}\!\upharpoonright_{\mathcal{E}_F} = \mathsf{CS}_F$.
$\mathcal{M}$ contains CS_F, therefore we have $c \in \mathrm{sub}(\mathsf{CS}_F) \subseteq \mathrm{sub}(\mathcal{B})$.
This is a contradiction with $c \in \mathrm{sub}(t) \setminus \mathrm{sub}(\mathcal{B})$ and shows that this case is impossible.

This shows $\mathcal{E}(\mathcal{B} \cup \mathsf{CS}) \cap \mathcal{E}_F \subseteq \mathcal{E}(\mathcal{B}) \cap \mathcal{E}_F$ and therefore $\mathcal{M}$ and $\mathcal{M}'$ agree on relevant evidence for F.
By Lemma 9.5, we conclude that $\mathcal{M}' \Vdash F$. □

Finally, with all steps of the atomization procedure in place, we can prove completeness w.r.t. atomic models.

Theorem 10.31 (Completeness for atomic models). *If* CS *is locally finite, logic* $\mathsf{JN}^V_{\mathsf{CS}}$ *is complete w.r.t. atomic* CS-*models, i.e.*

$$\Vdash^{\mathsf{CS}}_a F \quad \Rightarrow \quad \vdash_{\mathsf{JN}^V_{\mathsf{CS}}} F$$

Proof. Proof by contraposition: assume that $\nvdash_{\mathsf{JN}^V_{\mathsf{CS}}} F$.
By Theorem 8.12, it follows that $\nVdash^{\mathsf{CS}}_i F$.
Alternatively, there exists an intermediate CS-model $\mathcal{M}$ with $\mathcal{M} \nVdash F$.
Equivalently, we have $\mathcal{M} \Vdash \neg F$. Define $G := \neg F$.
Applying Lemma 10.22 to $\mathcal{M}$ and G, there exists a finite intermediate CS-closed model $\mathcal{M}_f$ containing CS_G such that $\mathcal{M}_f \Vdash G$.
Applying Lemma 10.28 to $\mathcal{M}_f$ and G, there exists a finite atomic CS-closed model $\mathcal{M}_a$ containing CS_G such that $\mathcal{M}_f \Vdash G$.
Finally, applying Lemma 10.30 to $\mathcal{M}_a$ and G, there exists an atomic CS-model $\mathcal{M}'$ such that $\mathcal{M}' \Vdash G$.
Equivalently, $\mathcal{M}' \Vdash \neg F$ and thus $\mathcal{M}' \nVdash F$ for an atomic CS-model $\mathcal{M}'$.
From this we conclude $\nVdash^{\mathsf{CS}}_a F$. □

11. Concluding remarks on nominals

With Theorem 10.31, we achieve the goal of having simple atomic model semantics for logic JN_V.

Logic JN_V is supposed to represent the state after updates with the fixed set of propositional variables V, with nominals corresponding to update terms. However, this is just a first step towards building a subscript-free variant of JUP.

JN_V lacks the ability to represent updates with more complex formulas, and has no notion of changing the nominal set.

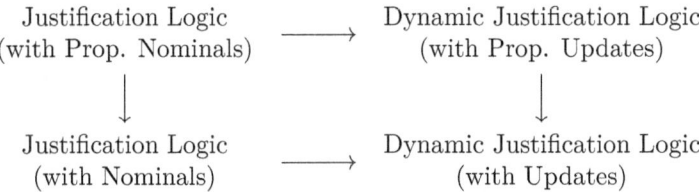

Figure 3: Roadmap for reconstructing updates

There are essentially two possible paths for extending JN_V to get back to the expressiveness of JUP, as shown in Figure 3.

It's possible to extend nominals to cover arbitrary formulas instead of just propositional nominals. Keeping atomic models in this case would require adding more axioms and restrictions on intermediate models that deal with "impossible" applications.

Alternatively, one can leave the nominals restricted to propositional variables, and attempt to add back the notion of updates with corresponding model dynamics. This seems to be a more natural extension.

11. Concluding remarks on nominals

One possible obstacle for extending $\mathsf{JN_V}$ to more closely reflect update dynamics is the fact that the set of "successful" updates is built into the language, specifically that V is fixed. $\mathsf{JN_V}$ in its current form cannot interpret update terms for updates that haven't happened yet. The corresponding way to extend $\mathsf{JN_V}$ would be to remove the use of the set V from the language itself and restrict it to the axiom system, which would then distinguish between nominals in the "successful" set and nominals that do not justify anything yet.

The local finiteness condition on CS is an unusual one: it puts an upper bound on possible constant specifications, when typical conditions like axiomatic appropriateness are, in a sense, a lower bound condition. This condition is sufficient to let the atomization procedure complete, but it's an open question whether it can be relaxed or dropped.

The atomization procedure itself may be useful in other justification logic systems. In fact, having nominals and explicit substitutions in $\mathsf{JN_V}$ made the procedure significantly more complex than the basic underlying idea of adding "fresh" evidence pairs to replace an irreducible application in the basis. Logical systems with generated models that do not have those features may enjoy a simpler version of the atomization procedure to obtain atomic model semantics.

Bibliography

[AGM85] Carlos E Alchourrón, Peter Gärdenfors, and David Makinson. On the logic of theory change: Partial meet contraction and revision functions. *The journal of symbolic logic*, 50(02):510–530, 1985.

[Art01] Sergei N Artemov. Explicit provability and constructive semantics. *Bulletin of Symbolic logic*, pages 1–36, 2001.

[Art06] Sergei N Artemov. Justified common knowledge. *Theoretical Computer Science*, 357(1):4–22, 2006.

[Art08] Sergei Artemov. The logic of justification. *The Review of Symbolic Logic*, 1(4):477–513, 2008.

[Art12] Sergei N Artemov. The ontology of justifications in the logical setting. *Studia Logica*, 100(1-2):17–30, 2012.

[BKR+10] Samuel Bucheli, Roman Kuznets, Bryan Renne, Joshua Sack, and Thomas Studer. Justified belief change. In *LogKCA-10, Proceedings of the Second ILCLI International Workshop on Logic and Philosophy of Knowledge, Communication and Action*, pages 135–155. University of the Basque Country Press, 2010.

[BKS11a] Samuel Bucheli, Roman Kuznets, and Thomas Studer. Justifications for common knowledge. *Journal of Applied Non-Classical Logics*, 21(1):35–60, 2011.

[BKS11b] Samuel Bucheli, Roman Kuznets, and Thomas Studer. Partial realization in dynamic justification logic. In *International Workshop on Logic, Language, Information, and Computation*, pages 35–51. Springer, 2011.

[BKS14] Samuel Bucheli, Roman Kuznets, and Thomas Studer. Realizing public announcements by justifications. *Journal of Computer and System Sciences*, 80(6):1046 – 1066, 2014. 18th Workshop on Logic, Language, Information and Computation (WoLLIC 2011).

[BRS12] Alexandru Baltag, Bryan Renne, and Sonja Smets. The logic of justified belief change, soft evidence and defeasible knowledge. In L. Ong and R. de Queiroz, editors, *Logic, Language, Information and Computation: 19th International Workshop, WoLLIC 2012, Buenos*

	Aires, Argentina, September 3–6, 2012. Proceedings, volume 7456 of Lecture Notes in Computer Science, pages 168–190, Buenos Aires, Argentina, 2012. Springer-Verlag Berlin Heidelberg.
[BRS14]	Alexandru Baltag, Bryan Renne, and Sonja Smets. The logic of justified belief, explicit knowledge, and conclusive evidence. *Annals of Pure and Applied Logic*, 165(1):49–81, 2014.
[BRS15]	Alexandru Baltag, Bryan Renne, and Sonja Smets. Revisable justified belief: Preliminary report. E-print, arXiv.org, March 2015. arXiv:1503.08141 [cs.LO].
[FH01]	Eduardo L Fermé and Sven Ove Hansson. Shielded contraction. In *Frontiers in Belief Revision*, pages 85–107. Springer, 2001.
[Fit05]	Melvin Fitting. The logic of proofs, semantically. *Annals of Pure and Applied Logic*, 132(1):1–25, 2005.
[Gär88]	Peter Gärdenfors. *Knowledge in flux: Modeling the dynamics of epistemic states*. The MIT press, 1988.
[KMOS15]	Ioannis Kokkinis, Petar Maksimović, Zoran Ognjanović, and Thomas Studer. First steps towards probabilistic justification logic. *Logic Journal of IGPL*, 23(4):662–687, 2015.
[KOS16]	Ioannis Kokkinis, Zoran Ognjanović, and Thomas Studer. Probabilistic justification logic. In *International Symposium on Logical Foundations of Computer Science*, pages 174–186. Springer, 2016.
[KS12]	Roman Kuznets and Thomas Studer. Justifications, ontology, and conservativity. *Advances in Modal Logic*, 9:437–458, 2012.
[KS13]	Roman Kuznets and Thomas Studer. Update as evidence: Belief expansion. In *International Symposium on Logical Foundations of Computer Science*, pages 266–279. Springer, 2013.
[KS16]	Roman Kuznets and Thomas Studer. Justification logic, October 2016. Lecture notes, Universität Bern.
[Ren08]	Bryan Renne. *Dynamic epistemic logic with justification*. ProQuest, 2008.
[Ren09]	Bryan Renne. Evidence elimination in multi-agent justification logic. In *Proceedings of the 12th Conference on Theoretical Aspects of Rationality and Knowledge*, pages 227–236. ACM, 2009.
[Stu13]	Thomas Studer. Decidability for some justification logics with negative introspection. *Journal of Symbolic Logic*, 78(2):388–402, 2013.
[Tar55]	Alfred Tarski. A lattice-theoretical fixpoint theorem and its applications. *Pacific journal of Mathematics*, 5(2):285–309, 1955.

Index

$\Box_{\mathcal{M}}$, 46
$[\Gamma^\circ]$, 17
$[\Gamma]$, 10, 17
$\Sigma \uparrow t$, 95
$\Sigma|_t$, 96
$\Vdash$, see truth relation
$\Vdash_{\mathsf{JUP_{CS}}}$, 14
$\Vdash_{\mathsf{JUP^\pm_{CS}}}$, 23
$\Vdash_a$, 71
$\Vdash^{\mathsf{CS}}_a$, 72
$\Vdash_i$, 71
$\Vdash^{\mathsf{CS}}_i$, 72
$\ominus_1$, 47
$\ominus_2$, 50
$\ominus_f$, 51
$\oplus$, 47
$\downarrow t$, 64
$\langle i, P, s \rangle$, 64
$\vdash_{\mathsf{JUP_{CS}}}$, 12
$\vdash_{\mathsf{JUP^\pm_{CS}}}$, 19
$\vdash_{\mathsf{JN^v_{CS}}}$, 67
$\wedge, \vee, \leftrightarrow$, 7
$t : F$, 10, 17, 64

ATm, 10, 18
$\mathcal{A}(\mathcal{M})$, 50
Atom_F, 102
acquired belief, 50

agreement
  on prop. valuation, 84
  on relevant evidence, 84
atomization, 113
  F-preserving, 102
  step, 112
axiom
  (App), 10, 18, 66
  (Init), 18
  (Int), 19
  (It), 11, 19
  (MC.1), 11, 18
  (MC.2), 11, 19
  (N_+), 66
  (N_-), 66
  (N_σ), 66
  ($\mathsf{N}_\triangleleft$), 66
  ($\mathsf{N}_\triangleright$), 66
  (Pers), 11, 18
  (Prop), 18, 66
  (Red.1), 10, 18
  (Red.2), 10, 18
  (Red.3), 10, 18
  (Roll), 19
  (Init), 11
  (Taut), 10
  (Up), 11, 18

$\mathcal{B}$, see basis

Index

$\mathcal{B}^{+\Gamma}$, 13, 22
$\mathcal{B}^{-\Gamma}$, 22
$\mathcal{B}_{(t,F)}$, 85
basis, 67
    atomic, 12, 20
    complexity, 92
    subterm set, 67
belief contraction
    closure, 51
    f-contraction, 51
    full, 50
    inclusion, 52
    naive, 47
    persistence, 50
    recovery, 48
    relative success, 50
    vacuity, 52
belief expansion, 47
belief set, 46
    induced, 46

CS, *see* constant specification
CS_F, 93
$\text{CS}\!\restriction_{\mathcal{B}}$, 92
$\text{CS}\!\restriction_c$, 92
$\text{Comb}(\Sigma_1, t, \Sigma_2)$, 98
$\text{cl}_{\mathcal{B}}$, *see* evidence closure
cmp
    $\text{cmp}(\mathcal{B})$, 92
    $\text{cmp}(\mathcal{M})$, 92
    $\text{cmp}(t)$, 65
closedness w.r.t. reasoning, 53
completeness
    JN_V
        atomic, 116
    $\text{JN}_\text{CS}^\text{V}$
        intermediate, 81
    JUP_CS, 14

$\text{JUP}_\text{CS}^\pm$, 41
consistent set, 30, 78
    maximal, 30, 78
constant specification, 11, 19, 67
    $\mathcal{B}$-fragment, 92
    c-fragment, 92
    axiomatically appr., 52
    $\text{JUP}_\text{CS}^\pm$-appropriate, 53
    locally finite, 94
    propositionally appr., 52
    relevant, 93

$\mathcal{E}(\mathcal{B})$, *see* evidence relation
$\mathcal{E}_F$, 83
$\mathcal{E}_X$, 103
evidence
    closure
        $\text{cl}_\mathcal{B}^{\text{JN}_\text{V}}$, 67
        $\text{cl}_\mathcal{B}^{\text{JUP}}$, 13
        $\text{cl}_\mathcal{B}^{\text{JUP}^\pm}$, 20
    pair, 8
    relation
        $\mathcal{E}^{\text{JUP}^\pm}$, 20
        $\mathcal{E}^{\text{JUP}}$, 13
        $\mathcal{E}^{\text{JN}_\text{V}}$, 67
    relevant, 83

$\text{Fin}(\mathcal{B}, F)$, 85
formula, 7
    $\text{Fm}_{\mathcal{L}_n^\text{V}}$, 63
    atomic, 10, 18
    evidence, 10, 17, 64
    $\text{Fm}_{\mathcal{L}_s^+}$, 10
    $\text{Fm}_{\mathcal{L}_s^\pm}$, 17
    rollback, 17
    subformula set, 83
    update, 10, 17
fresh variable, 99

Index

induction on buildup of $\mathcal{E}(\mathcal{B})$, 68

$\mathsf{JN_V}$, 66
$\mathsf{JN^V_{CS}}$, 67
JUP, 10
$\mathsf{JUP_{CS}}$, 11
$\mathsf{JUP^\pm}$, 18
$\mathsf{JUP^\pm_{CS}}$, 19

language
    $\mathcal{L}^V_n$, 63
    $\mathcal{L}^+_s$, 9
    $\mathcal{L}^\pm_s$, 17

$\mathcal{M}$, *see* model
$\mathcal{M}^{CS}$, 92
$\mathcal{M}^\circ$, 48
$\mathcal{M}^{+\Gamma}$, 13, 22
$\mathcal{M}^{-\Gamma}$, 22
$\mathcal{M}_\Phi$, 31, 78
model
    CS, 23
    CS-closed, 92
    CS-closure, 92
    agreement, 84
    almost finite, 83
    atomic, 71
    CS-model, 14, 72
    complexity, 92
    finite, 83
    induced, 31, 78
    initial, 14, 23
    initial part, 48
    intermediate, 71
    $\mathsf{JN_V}$, 70
    JUP, 13
    $\mathsf{JUP^\pm}$, 22
    reachable, 49

restriction
    (R.1), 71
    (R.2), 71
    (R.3), 71
    rolled-back, 22
    updated, 13, 22

n_P, 63
nominal, 59, 63

$PV(F)$, 99
Prop, 10, 17, 63
propositional valuation, 13, 70
    relevant, 83
propositional variable, 10, 17, 63

$\mathsf{rk}(F)$, 33, 80
$\mathsf{rk}_{Tm}(t)$, 32
rank
    formula, 33, 80
    term, 32
relevant
    constant specification, 93
    evidence, 83
    propositional valuation, 83
rule
    (AN), 11, 19
    (MP), 11, 19, 66

$\mathcal{S}(\mathcal{B}), \mathcal{S}(\mathcal{B})^*$, 94
$\mathsf{sub}(\mathcal{B})$, 67
$\mathsf{sub}(t)$, 63
$\mathsf{subf}(F)$, 83
soundness
    $\mathsf{JN^V_{CS}}$, 75
    $\mathsf{JUP_{CS}}$, 14
    $\mathsf{JUP^\pm_{CS}}$, 25
substitution, 94

Index

   basic, 64
   combining, 98
   filtering, 96
   lifting, 95
   set, 94

term, 7
   application, 9, 17, 63
   atomic, 10, 18
   complexity, 65
   constant, 9, 17, 63
   nominal, 63
   normal, 71
   subterm set, 63
   $\mathsf{Tm}_{\mathcal{L}_n^\mathsf{v}}$, 63
   $\mathsf{Tm}_{\mathcal{L}_s^+}$, 9
   $\mathsf{Tm}_{\mathcal{L}_s^\pm}$, 17
   update, 9, 17
   variable, 9, 17, 63
truth relation
   JN_V, 70
   JUP, 14
   $\mathsf{JUP}^\pm$, 22

$\mathcal{U}_\Gamma$, 13, 22
$\mathsf{up}(F)$, 9, 17
update set, 13, 22

v, *see* propositional valuation
v_F, 83
validity
   $\mathsf{JN}_{\mathsf{CS}}^{\mathsf{V}}$, 72
   $\mathsf{JUP}_{\mathsf{CS}}$, 14
   $\mathsf{JUP}_{\mathsf{CS}}^\pm$, 23

$X_{F,\mathcal{B}}$, 99

Erklärung

gemäss Art. 28 Abs. 2 RSL 05

Name/Vorname: Kashev Alexander
Matrikelnummer: 12-104-600
Studiengang: Informatik
Bachelor ☐   Master ☐   Dissertation ☒
Titel der Arbeit: Justification with Nominals

Leiter der Arbeit: Prof. Dr. Th. Studer

Ich erkläre hiermit, dass ich diese Arbeit selbständig verfasst und keine anderen als die angegebenen Quellen benutzt habe. Alle Stellen, die wörtlich oder sinngemäss aus Quellen entnommen wurden, habe ich als solche gekennzeichnet. Mir ist bekannt, dass andernfalls der Senat gemäss Artikel 36 Absatz 1 Buchstabe r des Gesetzes vom 5. September 1996 über die Universität zum Entzug des auf Grund dieser Arbeit verliehenen Titels berechtigt ist.

....
Ort/Datum   Unterschrift

Curriculum Vitae

Personal Details

Name	Alexander Kashev
Nationality	Russian
Date of Birth	March 30, 1987
Place of Birth	Moscow, USSR

Education

since 2012 — *Doctoral studies in Computer science*
Supervisor: Prof. Dr. Thomas Studer
Logic and Theory Group, Institut für Informatik, Universität Bern
Bern, Switzerland

2004—2009 — *Specialist (equivalent B.Sc. + M.Sc.) in Mathematics*
Faculty of Mechanics and Mathematics, Lomonosov Moscow State University
Moscow, Russia

2001—2004 — *Secondary education*
General Education Secondary School 54
Moscow, Russia

1994—2001 — *Elementary and secondary education*
General Education Secondary School 1215
Moscow, Russia

www.ingramcontent.com/pod-product-compliance
Lightning Source LLC
Chambersburg PA
CBHW070234180526
45158CB00001BA/495